The Causal Universe

The Causal Universe

edited by

GEORGE F.R. ELLIS

MICHAEL HELLER

TADEUSZ PABJAN

Copernicus Center PRESS

Editing:
Aeddan Shaw

Cover design:
Mariusz Banachowicz

Layout:
Mirosław Krzyszkowski

Typesetting:
MELES-DESIGN

Publication Supported by the John Templeton Foundation Grant
"The Limits of Scientific Explanation"

ISBN 978-83-7886-034-1

Kraków 2013

Copernicus Center PRESS

Publisher: Copernicus Center Press Sp. z o.o.,
pl. Szczepański 8, 31-011 Kraków,
tel/fax (+48) 12 430 63 00
e-mail: marketing@ccpress.pl
www.ccpress.pl

Table of Contents

Part II
Ultimate Causality

The nature of causation is a core issue for science, which can be regarded as the move from a demon-centered world to a world based on reliable cause and effect, tested by experimental verification.

(...) Physics is the basic science, characterized by mathematical descriptions that allow predictions of physical behavior to astonishing accuracy and underlies the other sciences. The key question is whether other forms of causation such as those investigated in biology, psychology, and the social sciences are genuinely effective, or are they rather all epiphenomena grounded in purely physical causation?

(...) I will claim here that there are indeed other types of causation at work in the real world, described quite well by Aristotle's four types of causes. There are of course many contexts in which different kinds of causality are experienced: in physics and chemistry, where particles and forces interact in a way described by variational principles and symmetries; in biochemistry and cell biology, where information is important and adaptation takes place; in zoology, where purpose, planning, and anticipation are important; and in psychology and sociology, where analytic reflection, symbolic understanding, values and meaning all are causally effective.

George Ellis, *On the Nature of Causation in Complex Systems*

Introduction

Physics is heavily marked by its philosophical origins. It is enough to relax methodological control and the physicist almost immediately slips into philosophical speculation. The only scientists who are immune to this danger are those who never go beyond their calculations or laboratory measurements. This phenomenon is, at least in part, due to the fact that many concepts which are of crucial importance for physics have been transported from philosophy to physics and still preserve a propensity towards their family home. One can speak about a process of the *migration of concepts* from philosophy to physics (or more generally to science) and quote a host of its examples. Concepts like: space, time, force and mass first come to mind.

But concepts never appear in isolation from other concepts, and are always situated within certain problems. And when they migrate to the scientific environment, they take with them all of this baggage. Of course, when they are already well placed in their new context, they are clothed in new connotations and a new conceptual network. Moreover, quite often a given concept goes back to philosophy, and when it does so, it never returns to exactly the same place as before (certainly, no concept 'ever steps in the same river twice'). This is one of the ways of enriching philosophical inquiries and opening them up to new perspectives.

The concept of causality belongs to the family of those selected concepts that have covered the distance between philosophy and physics many times. This makes this process a fascinating subject matter for interdisciplinary debate, and this is why, when George Ellis suggested the 'Causal Universe' as the title for this volume, the idea seemed so obviously correct that there were no counterproposals. The volume is the outcome of the 16th Cracow Methodological Conference 'The Causal Universe', held in Cracow, Poland, on 17–18th of May 2012, with George Ellis as its Honorary Guest. The chapters contained in this book have evolved from the contributions presented by their authors during the conference.

The manifold migration of the causality concept between philosophy and physics is beautifully reflected in the content of the volume. It is divided into three parts. The first part, TOP DOWN CAUSALITY AND COMPLEXITY, initiates the discussion around two of the main notions which reappear time and again in many contributions. The second part, CAUSALITY AND THE STRUCTURE OF THE UNIVERSE, transfers the discussion into the cosmological context. It culminates in the third part, ULTIMATE CAUSALITY, where we move to the philosophical and theological aspects of the problem. Although we go beyond the realm of science, we try not to lose our scientific perspective.

PART ONE. George Ellis in his opening chapter ('Why Are the Laws of Nature as They Are? What Underlies Their Existence?') not only sets the stage for the entire volume, but also draws the main lines of the plot that unfolds in it. The laws of physics define the possibility space within which the universe and everything in it come into being. If we aspire to understand the universe and ourselves within it, we cannot avoid inquiring as to what underlies the existence of the laws of physics. If we take the existence of the laws of physics for granted, as is usually done in many otherwise penetrating discussions, our understanding cannot be complete. We are living in an evolving Universe that was initially structureless. Structures emerge from the underlying basis of physical laws. Complexity arises in modular hierarchical structures in manifold ways. No biological evolutionary processes of any kind would be possible if physical laws were substantially different. Is the ultimate reason underlying the lowest-level laws on which everything else is based pure chance, probability, necessity, or purpose? Although neither science nor philosophy can give a definite answer, Ellis argues that if one wants to relate one's understanding to the deeper meaning of personal life, the last option is the most attractive one.

The second of Ellis' papers ('Top-Down Causation as the Key to the Emergence of Complexity') has a more technical character. He quotes persuasive arguments for the existence of top down causality and its crucial role in the emergence of complexity. A key assumption underlying most contemporary scientific work is that causation is bottom up all the way: particle physics underlies nuclear physics, nuclear physics underlies atomic physics, atomic physics underlies chemistry, and so on and, consequently, all the higher level structures are at least in principle reducible to particle physics. Ellis argues that this view is wrong in general and, in particular, in the case of physics. The top-down form of causation does not occur by violating physical laws but, on the contrary, it occurs through the laws of physics, by setting constraints on lower level interactions. He also claims that the emergence of true complexity is not possible through bottom up effects alone; it requires a reversal of information flow from bottom-up to top-down.

In his chapter ('Models of the Cosmos and the Emergence of Complexity') Jean-Philippe Uzan places Ellis' analysis of top-down causality and the growth of complexity in the context of the contemporary standard cosmological model. To do so, he first discusses hypotheses underlying the physical representation of the universe and observational data as potential tests of these hypotheses. This shows a deep connection between theoretical physics and cosmology. The cosmological model provides a context for the emergence of complexity. The universe is the largest embedding structure in which all the phenomena of lower complexity levels develop. It is thus important to understand how the structure of the universe and its evolution offers the possibility for the emergence of other levels. In this cosmological context, the role of effective phenomena and the concept of causality are central. Uzan underlines what he calls the scale decoupling principle; it refers to the fact that there exist energy scales below which effective theories are sufficient to understand a given set of physical phenomena. They give a successful explanation at a given level of complexity based on concepts of that particular level. For instance, we do not need to understand string theory in order to formulate nuclear physics and we do not need to know anything about nuclear physics to develop atomic physics or chemistry. It is also interesting to notice that the appearance of levels of complexity is localized both in time and space, so that the region of the universe in which top-down causation is efficient is limited. This remains in agreement with the usual relativistic causality; however, the typical propagation speed of the action of higher levels on lower levels is much smaller than the speed of light, so that the future light cone of influence is expected to be very narrow.

Derek Raine ('Causality and Complexity') plays the role of an *advocatus diaboli* and argues that the arguments on behalf of top down causality are unconvincing. In linear systems there is a direct link between cause and effect. This is not the case in non-linear systems where causes are inter-related. For instance, what was the cause of the First World War? This is clearly an unanswerable question. The difficulty in assigning a cause arises because the situation is non-linear: that is, there is feedback. In a linear system if we add causes, we must add their effects: if *a* causes *A* and *b* causes B, then *a* and *b* acting together cause the outcomes *A* and *B* together. Once we introduce interactions, hence non-linearity, either between different parts of a system or self-interaction, this simple link between cause and effect is lost. After discussing several examples of apparent top down causality (some of these examples are quite technical), Raine concludes that 'there is no denying that there is more to causality than the Aristotelian 'efficient cause' (the sculptor's chisel on the stone)'. In Raine's opinion, supporters of top-down causality are well aware of all of this. However, they are not satisfied with calling it adaptive evolution but call it top-down causality. He remarks that 'calling things by different names is harmless except when it isn't: in particular when it leads to category errors such as the attribution of existence to the thing from the existence of the idea of the thing'.

PART TWO. In this part, by the universe we understand not only the cosmos as it is investigated by cosmology, but also all its substructures as investigated by other physical theories. We start with quantum mechanics, general relativity and astrophysics, and then go to more exotic research areas: physical theories based on noncommutative geometry and the multiverse idea.

Very interesting, and well experimentally documented, things related to causality happen in quantum mechanics. Any discussion of causality would remain incomplete without even a brief presentation of the problems of causal relations and causal explanations specific to quantum mechanics. This task has been taken up by Marek Kuś ('The Uncertain Future and the Ambiguous Past in Classical, Quantum and General Non-signaling Settings'). With only a slight exaggeration, one can say that physics is the art of determining correlations between various parts of the physical world. Kuś gives a short but very lucid presentation of some aspects of this topic exhibited by recent developments in the theory of quantum correlations. Local theories, admitting only local correlations put some limits on possible correlations among results of measurements called Bell inequalities. Their violation, confirmed experimentally, testifies to the nonlocal character of the quantum world, and should be regarded as proof of the nondeterministic nature of quantum mechanical reality and its intrinsic, ontological randomness. In the classical setting, uncertainty about future events had an epistemological character – an event is random for us due to our lack of knowledge about all of its causes, whereas in quantum physics the future seems ontologically undetermined.

The issue of causality and the emergence of structures is strictly linked with the problem of reductionism. If its strong version is true, we should exclude top-down causality and reduce complex structures to their simplest elements. The problem, as it appears in general relativity, is considered by Julian Barbour ('Reductionist Doubts'). He notices that if general relativity implements Mach's principle, reductionism is challenged. Mach's principle asserts that parts of the universe and their interactions do not determine the whole; on the contrary, the whole determines the way the parts behave. Whether general relativity implements Mach's principle has been a matter of controversy every since Einstein created his theory of gravitation. Barbour has given what he believes is the correct definition of Mach's principle and proposes an alternative interpretation of general relativity in which this theory appears much more holistic than in the usual interpretation. The Machian interpretation of general relativity radically changes the way we conceive of the parts and not just the way they interact. At the end, Barbour discusses the possible implications of a holistic quantum view of the universe.

In the last hundred years, many branches of astrophysics have contributed to the advancement of cosmology. For instance, the hot beginnings of the Universe were confirmed observationally in the 1960's with the discovery of cosmic microwave background (CMB) radiation. Investigation of CMB, in turn, has allowed us to develop the Big Bang theory further. The X-ray background (XRB) constitutes another section of quasi-isotropic radiation flux reaching the

Earth. The X-ray window was opened for astrophysics 50 years ago. The goal of Andrzej M. Sołtan's paper ('X-ray Background and Cosmology') is to inform us how this window has contributed to our understanding of the universe. One of the revolutionary turns which changed our perception of galaxies as the basic building blocks in the Universe, was due to X-ray observations of quasars and Seyfert galaxies and the activity of their nuclei. These two classes of extragalactic objects were known of earlier, but neither optical nor radio data provided an inkling that supermassive black holes are responsible for virtually all kinds of the nuclear activity of galaxies. Clusters of galaxies were the first extragalactic objects which were clearly recognized as X-ray sources. The detection of X-ray emissions from clusters bolstered the idea of dark non-baryonic matter as a dominating component of the matter density of the Universe. X-ray observations of hot intracluster gas allowed for the quantitative investigation of a gravitational field within clusters and the common acceptance of the reality of dark matter. Until recently, the question of 'dark matter' was accompanied by similar one of 'dark baryons'. This term was used to label a discrepancy between the density estimates of the baryonic matter in the early Universe and the present-day observations in the local Universe. X-ray observations have also alleviated this problem. It was possible because in soft X-rays, below 1 keV, a fraction of 'missing' baryons are not completely 'dark'.

With the chapter by Andrzej Sitarz ('Causality and Noncommutativity') we turn from the world of quasars and black holes to the most fundamental level of physics. In contemporary physics reigns a certain schizophrenic duality. On one hand, we have gauge theories, which nicely allow the formulation of classical and perturbative quantum field theory; on the other hand, we have a purely geometrical theory of gravity, mathematically set in the Riemannian (or pseudo-Riemannian) geometry that gives no hints as to its quantum version, the mythical quantum gravity. So far, the attempts have been directed towards the modifications of general relativity in order to assimilate it into a gauge theory. An attempt, stemming from the works of Alain Connes, consists in looking in the opposite direction. Instead of regarding gravity as some sort of gauge theory, we should rather consider gauge theories as some modifications of pure gravity. Such a modification proposed by Connes with this aim is the so-called noncommutative geometry. The models of noncommutative geometry pose a big challenge to the standard concepts of locality and causality. Can we formulate a version of causality which would correspond to the deformed symmetries? Most of our intuitions and approaches, also in the quantum field theory, are based on the notion of time as a continuous variable. What can we do if time is supposed to be a noncommutative variable? Can one allow at all for noncommutative time? Sitarz offers us a thorough discussion of these fundamental questions.

Michael Heller's paper ('Bottom-Up Causality in a New Setting') is, in a sense, complementary with respect to that of Sitarz. Heller tries to reconstruct an 'ontology of the Planck level' based on the structural properties presupposed or suggested by noncommutative geometry. Such an ontology, which could be timeless and nonlocal, is drastically different from the ontologies presupposed by classical physics. In particular, the concept of causality should be modified accordingly. If we agree that causality is more than a simple sequence of events then it is just the dynamics that gives a causal nexus between cause and effect. In the noncommutative setting we are confronted with a generalized, nonlocal sort of dynamics. We should forget about 'the cause here and the effect there' since the concepts 'here' and 'there' are not available. They are totally engulfed by the global dependencies between various aspects of the physical world. If we agree that the main characteristics of top-down causality is a holistic type of explanation, and the main characteristic of bottom-up causality is a reductionist type of explanation, then what happens on the 'bottom', i.e., on the Planck level, looks more like a top-down than a bottom-up causality: global properties are not reduced to local properties but, vice versa, local properties are derived from global ones. It is rather a top-down than a bottom-up causality. Heller speaks here about a top-down causation with the reversed top-down arrow (pointing from 'down' to 'up').

We have just considered the most fundamental and the 'smallest' level of physics; at the other end, as it were, of the scale of magnitude there is (if it is) what is called the multiverse. It is considered by Mariusz P. Dąbrowski ('Varying Physical Constant Cosmologies and the Anthropic Principles'). He discusses the roots of the idea of cosmologies with varying physical constants and its realization in specific models of particle physics. He also considers some examples of anthropic coincidences which limit the variability of physical constants in our piece of the universe. In this context, the idea of the multiverse appears in a natural way. The discussion is underpinned by questions such as: Why are the physical constants which we measure of the values that they are? Do these constants change in time and space? How different would our world have been if the constants had had different values from what they have in our corner of the universe? Are there any regions of space in which physics does not allow for our type of life? And, in the theological context, we may ask whether there was initially any freedom in choosing physical constants, and why they have been chosen in a way they are now. Dąbrowski argues that, philosophically, the idea of varying fundamental constants is very attractive since it gives an opportunity for considering all of the possible options of the evolution of the universe which are logically possible and mathematically admissible.

PART THREE. The problem of causality is a typical problem that has migrated from philosophy to physics, and any discussion of this problem would be painfully incomplete without turning to the great metaphysical issues. We start, by the way of introduction, with a glance at the origins of the concept of causality.

All important philosophical questions have their origins in Greek philosophical thought. It is also true as far as the causality issue is concerned. Bogdan Dembiński ('Causality Issues in Ancient Greek Philosophy') gives us an in-depth overview of the Greek concepts of causality in which he goes beyond the 'common knowledge' on this subject. According to the Greeks, the decisive role in creating the form of the visible world was played by a system of limits which determine the cosmic order and harmony. This structure is shaped by the interactions and relations among its constituent elements. They are expressed by proportions, arrangements and locations and, consequently, the understanding of the cosmic order is strictly connected with its mathematical description. This approach reached its full development in Plato's Academy (in the works of Eudoksos of Knidos, Speuzipos, Xenokrates and Herakleidos Pontikos). But what is the cause of the existence of limits, their form and multiplicity? The Greeks found the explanation in the theory of the Supreme Principles of Being. These Principles are: the One, One and Many, the Opposites, the Limit and the Limitlessness, Love and Hatred, the Cosmic Intellect (Logos), the Divine Mind or the Divine Organizer of the World (Demiurg). The Principles constitute the end of any justification; one cannot go beyond them. To uncover the truth is to reveal the implicit cosmic order. Only the intellect can accomplish it and it does so because it is itself a part of this order.

Having prepared a general stage, we proceed to ask essentially the same question in the contemporary setting. William R. Stoeger, S.J. ('Cosmology, Evolution, Causality and Creation: the Limits, Compatibility and Cooperation of Scientific and Philosophical Methodologies') provides an integrated sketch of the causality problem as it reveals itself in both our scientific image of the world and from our philosophical perspective. He also discusses different types and models of causality, and their relationships with one another. Then, after briefly reflecting on what we know of the Planck era and quantum cosmology, he explores a different kind of 'causality' which takes him beyond the natural sciences into the realm of philosophy. This is causality, often called 'primary causality', regarded as the ultimate source of existence and order. It is 'causality beyond causality' in the sense that it does not provide an alternative to physics or the other natural sciences, but rather gives the grounds for all other causes and their existence.

Thomas Tracy ('God and the Causal Structures of Nature: Some Puzzles') notices a strange coincidence: 'talk about God has always been an exercise in the art of thinking about a subject matter that we know from the outset must exceed our comprehension' and 'contemporary physics… too leads us to the boundaries of what we can conceive'. His aim is to consider the impact of some concepts from contemporary physics, especially quantum theory, on thinking about God's relation to the natural order. Crucial to this relation is the concept of creation, but the notion of 'causing being' should only, in this context, be understood as a metaphysical black box; the most we can do is to indicate the effect of the divine action. The idea of God as the Cause of Created Causes introduces the familiar scheme of primary and secondary causation. Creatures operate as causes solely by inducing changes in other things. One consequence of this view is that, *contra* intelligent design theory, we need not insist on the incompleteness of biological explanations in order to affirm that God's purposes are at work in the history of life. The properties of quantum systems, such as indeterminateness, nonlocality and underdetermination seems to disrupt the classical theological scheme of divine action through secondary causes. The quantum world appears itself as an evolving structure of potentiality. The question arises: what is God's relation to this underdetermined world? God acts as its creator, just as with every entity and event in the history of the universe, but the puzzle is how to conceive of *what* God creates. The author enters into all these questions with a responsible knowledge of modern physics and a profound theological expertise. His 'modest conclusion' is: 'we theologians may wish to keep using the primary/secondary causation scheme, but if we do so, we ought to recognize and grapple with the fact that its metaphysical foundations have been undermined, and it is not clear how to provide a new basis for a comparable way of relating divine and created causes'. He finds some comfort in the statement 'that theology is not alone in finding that old ways of thinking about causality have come unstuck in the encounter with quantum mechanics'.

Willem B. Drees ('God as a Ground? Cosmology and Non-Causal Conceptions of the Divine') continues Tracy's line of thought. He takes into account the possibility that a 'first cause' view of God may not be integrated with cosmic natural history but nevertheless, in his view, the mystery of existence remains. After an excursion into some issues of natural theology in their newer incarnations, and those of Big Bang theory and its limitations, Drees embarks on his main topic. The development of current speculations, which go 'beyond the Big Bang theory', gives him a context to reflect further upon the idea of a *first* cause. With the development of new scientific theories, our knowledge of the world is not merely enlarged, but rather restructured. Entities and structures postulated by new theories had not been envisaged before. New theories lead to a reinterpretation of the world. This requires the rethinking of many philosophical and theological issues, the first cause problem included. One might envisage the end of a particular form of the cosmological argument, but the fundamental question remains: why does there exist something rather than nothing? Perhaps, the transcendent is not to be thought of in temporal terms. Given the role of mathematics, and its independence from the physical dimensions of time and space, one might try to draw on mathematics and logic to imagine 'transcendence'. In mathematics, axioms are not so much the cause of the theorems, but rather a formal basis for all subsequent theorems. Is it a return to Anselm's ontological argument which also argued for God along the lines of logic rather than of causality? This opens a perspective for reflecting upon a different way of conceptualizing God, as a 'ground' rather than as a 'cause'. In the final section, Drees discusses the nature of theology as a particular type of human construction.

Why does the universe exist and why is it as it is? The reading of this book offers a unique opportunity to broaden our understanding of how to combine our vast knowledge of the laws governing the universe with responsible philosophical inquiry in the quest for the ultimate explanation of the world's existence and its specific nature.

Michael Heller

Part I

Top-Down Causality and Complexity

George F.R. Ellis
University of Cape Town

Why Are the Laws of Nature as They Are? What Underlies Their Existence?

1. The Existence of the Laws of Nature

A deep issue in both cosmology and human life is, *What underlies the existence of the laws of nature?* These laws define the possibility space within which the Universe and life come into being. We want to understand the existence and nature of causal laws that allow true complexity, such as ourselves – our bodies and minds, our actions and thoughts and emotions – to come into being. These laws are simply taken for granted in most biological and evolutionary discussions; but in the context of fundamental discussions concerning cosmology and life, one is entitled to query both their nature and their existence, for they shape the nature of physical existence and are the foundational reasons that any life whatever is possible. No life, and no biological evolutionary processes of any kind, would be possible if these laws were substantially different.

The fundamental laws of nature are based in mathematically describable variational principles for physical entities[1] invariant under symmetry groups and subject to quantum mechanical dynamics (Cottingham and Greenwood 2007). The coupling constants that relate

[1] One considers a quantity (the 'action'), depending on the path from an initial to final point, and determines on which path this quantity is either a minimum or maximum as one varies the path.

resulting forces and particles are such that complexity, including consciousness, can emerge out of these basic physical constituents; this is a priori a highly improbable state of affairs (Davies 1982, Barrow and Tipler 1986, Balashov 1991, Rees 2000, 2001).

A series of fundamental issues arise:

- Why do any such laws exist at all, and why do they have the nature they have, leading to our physical and mental existence?
- Is the ultimate reason pure happenstance, probability, necessity, or purpose?
- What is the nature of their existence – is it prescriptive or descriptive?

2. The Hierarchy of Complexity

Our cosmological context (Harrison 2000, Silk 2005, Ellis 2006) is an evolving Universe that is initially structureless, but eventually physical processes lead to the existence of galaxies, stars, and planets. The emergence of true complexity, including living beings such as ourselves, occurs on some suitable planets imbedded in this larger context. Complexity arises in modular hierarchical structures (Simon 1982, Peacocke 1989, Flood and Carson 1990, Booch 1994, Scott 1995, Campbell and Reece 2005, Ellis 2008b), which underlie biological function, as indicated in Table 1. These structures emerge from the underlying physical basis in 3 very different, but interrelated ways:

Table 1

The Hierarchy of Structure and Complexity Underlying Human Existence is Characterized by the Various Sciences Appropriate for Studying Each Level

Sociology/Economics/Politics
Psychology
Botany/Zoology/Physiology
Cell Biology
Biochemistry
Physical Chemistry
Atomic Physics
Nuclear Physics
Particle Physics

- *In evolutionary terms*, on very long timescales ($>10^6$ years). There was no life in the Universe when it was 300,000 years old or on Earth 4 billion years ago; living beings came into existence through an evolutionary process that lasted billions of years and generated complexity where none had existed before.
- *In developmental terms*, on medium timescales (hours to 10^2 years). Each higher-level animal, including human beings, starts off as a single living cell and then develops into a multicellular organism ($\sim 10^{13}$ interacting cells) through developmental processes that create functioning physiological structures (Gilbert 2006).
- *In functional terms*, on short timescales (milliseconds to minutes). We are each made up of inanimate objects (protons, neutrons, and electrons) that are fashioned into physiological systems that together make a living being. Life emerges through the ongoing cooperation of these basic physical entities (Campbell and Reece 2005).

Once life has been initiated, the developing organisms must continually keep functioning effectively at each stage of the developmental and evolutionary processes, despite the enormous changes in structure and complexity that occur as complexity develops. Thus, the functional processes evolve on both evolutionary and developmental timescales.

I suggest that at each level of the hierarchy of complexity, universal principles apply, describable as effective laws applicable at that level. Each level exists in its own right, even though it is based in lower levels. The way the lower-level laws work out is shaped by the higher-level contexts in which they act, deriving from action at lower level, but with their own autonomy of operation, leading to effective laws at each level. These effective laws control what happens at each level in a way independent of time and place and independent of our understandings and descriptions, hence they may be thought of as having an existential reality describable in Platonic terms[2]. This is all possible because both bottom-up and top-down causation takes place in the hierarchy of complexity.

Bottom-up Action

A fundamental understanding we have attained through 300 years of scientific endeavor is that all complex objects are made of the same basic materials (atoms fashioned into molecules), with micro-forces determining what happens at the lower levels, and hence underlying properties at the higher levels. *Bottom-up action* is when what happens at the higher levels in the hierarchy is controlled by what happens at the lower levels (Fig. 1, left). Examples are abundant:

---

[2] A Platonic kind of existence is an existence in an abstract space, separate from physical reality.

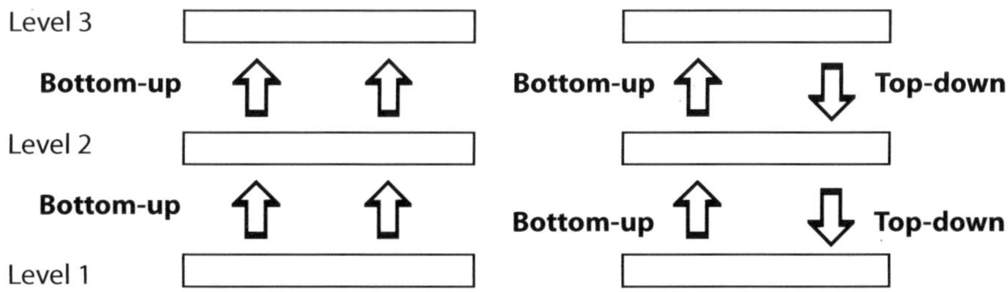

Figure 1 *Left*: Bottom-up causation. *Right*: Bottom-up and top-down causation. The fundamental importance of top-down causation is that it changes the causal relation between upper and lower levels in the hierarchy of structure and organization.

- Microphysics underlies macrophysics, e.g., the kinetic theory of gases and the theory of solids, determining properties of gases (such as the temperature-pressure relation) and solids (such as electrical conduction and thermal capacity) from the underlying material properties (Goodstein 1985, Durrant 2000).
- Physics underlies chemistry, e.g., determining thermodynamic properties of matter and the nature of the chemical bond (Laidler et al. 2003).
- Molecular processes underlie organic life (Scott 1989).
- Cells with their own internal function underlie all life (Harold 2001).
- Physics and chemistry underlie the functioning of the brain (Scott 1995).
- Individual human behavior underlies the functioning of society (Berger 1963).

Bottom-up causation is the focus of reductionist accounts of life. However, although it is indeed a key factor in what occurs, it is not the whole story.

Top-down Causation

The fundamental importance of top-down causation is that it changes the causal relation between upper and lower levels in the hierarchy of structure and organization: causes come from some upper levels to lower levels, as well as from lower levels to upper levels (Ellis 2008a,b).

In addition, top-down causation takes place in the hierarchy of complexity (Campbell 1974, Murphy and Ellis 1995, Noble 2007, Ellis 2008b). Through boundary effects (linking the system to the environment) and structural relations in the system itself, the higher levels of the hierarchy causally effect what happens at the lower levels (Fig. 1, right) by determining the context of the lower-level processes. Through coordinating the action of the lower levels, the higher levels attain their causal effectiveness. Top-down causation is prevalent in the real physical world and in biology because no real physical or biological system is isolated. It is the occurrence of multiple topdown actions in conjunction with bottom-up action that enables the self-organization of complex systems. Here are some examples:

1. The synthesis of light elements in the early Universe (Silk 2005). The amount of helium produced depends on the rate of change of temperature in the expanding Universe, which is controlled by the gravitational equations and the average amount of matter in the Universe. Thus, quantities defined at the cosmological level control the products of detailed nuclear reactions at the micro level.
2. The functioning of the molecular systems underlying cardiac behavior is determined in a top-down way by the state of the heart as a whole (Noble 2007).
3. Training of artificial neural networks to perform a specific task (say, letter recognition) determines the interaction weights in the network (Bishop 1999). This is a form of top-down causation from the pattern to be recognized (a high-level concept, as it is defined in terms of the relation between the elements) to the low-level property of network weights. Decision making is a property of the network rather than of any single cell.

4. The power of the human mind in the real world: for example, design and construction of a jumbo jet aircraft, thereby determining the disposition of myriads of atoms (Ellis 2008b).

Numerous other examples are given in Ellis (2008b), which also looks at the key question: *How is effective top-down causation possible without violating the integrity of bottom-level causation?* The basic answer is that the lower-level physics does not uniquely determine higher-level outcomes, not only because of statistical effects, but because quantum theory does not determine a unique physical outcome, *even in principle* (Feynman 1985, Penrose 1989b, 2005, Polkinghorne 2002, Al-Khalili 2003). Physics gives the statistics of outcomes, but not the unique actual outcome. This opens up the space for selectional principles to operate as a form of top-down action.

The Causal Efficacy of Nonmaterial Entities

The overall importance of this discussion is that it is through top-down action that immaterial entities can have causal effects on the physical world (Ellis 2008b).

The first important example is social constructions, such as the rules of chess and the value of money. These are indeed causally effective in the real world, but they are not the same as a state of any single physical object; in particular, they are not equivalent to any individual's brain state (although they are realized through such states).

First, the ontological status of rules, such as, say, for chess, is not dependent on any individual's brain or state of mind. Rather, such rules are an abstract social construction, shared by many people and arrived at through social interactions over the course of time, with many different embodiments: they are written down, can be talked about, are thought about, are embodied in computer programs, and so on. They are not physical entities, but they are causally effective. Similarly, physically, money is just coins or pieces of paper with patterned marks on them. This does not explain its causal significance. The effectiveness of money, which can cause physical change in the world such as the construction of buildings, roads, bridges, and so on, by top-down action of the mind to material objects, is based in social agreements that lead to the value of money (pricing systems) and exchange rates. These are abstract entities arising from social interaction over an extended period of time and are neither the same as individual brain states nor equivalent to an aggregate of current values of any lower-level variables (although they may be represented by, and are causally effective through, such states and variables).

Second, abstract entities of a Platonic nature can also be causally effective in the real world through the action of the human mind. For example, mathematics comprehension and utilization is a case of causation from a Platonic world of mathematical abstractions to the human mind, being realized in details of neuronal connections and then into the real world, where it is causally effective in terms of creating patterns on paper and through underlying physics, engineering, commerce, and general planning.

The existence of a Platonic world of mathematical objects is strongly argued by Penrose (1997, 2005) and Changeux and Connes (1998), the point being that major parts of mathematics are discovered, rather than invented (rational numbers, 0, irrational numbers, and the Mandelbrot set being classic examples; see, e.g., Seife 2000 for the case of 0). They are not determined by physical experiment, but are rather arrived at by mathematical investigation. They have an abstract, rather than an embodied, character; the same abstract quantity can be represented and embodied in many symbolic and physical ways (Penrose 2005), and these representations form an equivalence class. They are independent of the existence and culture of human beings; it is plausible that the same features will be discovered by intelligent beings in the Andromeda Galaxy as here, once their mathematical understanding is advanced enough (which is why they are advocated as the basis for interstellar communication). This Platonic world is, to some degree, discovered by humans and represented by our mathematical theories; that representation is a cultural construct, but the underlying mathematical features they represent are not – indeed, like physical laws, they are often unwillingly discovered, as with, for example, the irrationality of the $\sqrt{2}$ and the number π. This world is causally efficacious in terms of the process of discovery and description: one can, for example, print graphic versions of the Mandelbrot set in a book, resulting in a physical embodiment in the ink printed on the page. The causal variables here are not coarse-grained lower-level variables and exist independent of the mind, even though they are discovered and comprehended there.

3. The Nature of Existence

It is the existence of these causally effective immaterial entities that enables us, indeed requires us, to contemplate kinds of existence other than that of merely physical entities. In discussing this, I *take as given the reality of the everyday world* – tables and chairs, and the people who perceive them – and then assign a reality additionally to each kind of entity that can have a demonstrable causal effect on that everyday reality. The problem then is to characterize the various kinds of independent reality that may exist in this sense. Taking into account the causal efficacy of all the entities discussed above, I suggest as a possible completion of the proposals by Popper and Eccles (1977) and Penrose (1997, 2005) that the 4 Worlds indicated in Table 2 are ontologically real (that is, they really do exist physically)[3]. These are not different causal levels within the same kind of existence; rather, they are quite different kinds of existence, but related to one another through causal links. The challenge is to show first that each is indeed ontologically real, and second that each is sufficiently and clearly different from the others that it should be considered as separate from them. I now discuss them in turn; see Ellis (2004) for a more detailed discussion.

[3] 'Ontology' is the philosophical study of the nature of being or existence.

Matter and Forces

World 1 is the physical world of energy and matter, hierarchically structured to form lower and higher causal levels whose entities are all ontologically real.

This is the basic world of matter and interactions between matter, based at the micro-level on elementary particles and fundamental forces, and providing the ground of physical existence. It comprises *inanimate objects* (both naturally occurring and manufactured), *living things* (amoebae, plants, insects, animals, etc.), and *human beings*, who have the unique property of self-consciousness.

The hierarchical structure in matter is a real physical structuration and is additional to the physical constituents that make up the systems themselves. It provides the basis for higher levels of order and phenomenology, and hence of ontology. Ontological reality exists at each level of the hierarchy. Thus, we explicitly recognize as being real the following: quarks, electrons, neutrinos, rocks, tables, chairs, apples, humans, the world, stars, galaxies, and so on. The fact that each is made up of lower-level entities does not undermine its status as existing in its own right (Sellars 1932). We can attain and confirm high representational accuracy and predictive ability for quantities and relations at higher levels, independent of our level of knowledge of interactions at lower levels, giving well-validated and reliable descriptions at higher levels accurately describing the various levels of emergent nonreducible properties and meanings.

Consciousness

World 2 is the world of individual and communal consciousness: ideas, emotions, and social constructions. This again is ontologically real (it is clear that these all exist) and causally effective.

These worlds are different from the world of material things and are realized through the human mind and society. They are not brain states, although they can be represented as such, for they do not reside exclusively in any particular individual mind. They are the foundation for social interactions and intellectual activity. They are causally effective because, as discussed above, they do indeed have the capacity to change the state of the physical world. The existence of houses and automobiles, of books and computers, of cities and factories, is evidence of this effectiveness.

Table 2

The Different Kinds of Reality Implied by Causal Relationships can be Characterized in Terms of 4 Worlds, Each Representing a Different Kind of Existence

World 1:	Matter and Forces
World 2:	Consciousness
World 3:	Physical and Biological Possibilities
World 4:	Abstract (Platonic) Reality

Physical and Biological Possibilities

World 3 is the world of possibilities. This characterizes the set of all physical and biological possibilities, from which the specific instances of what actually happens in World 1 are drawn.

This world of possibilities is ontologically real because of its rigorous prescription of the boundaries of what is possible – it provides the framework within which World 1 exists and operates, and in that sense it is causally effective. This world is different from the world of material things, for it provides the background within which that world exists. In a sense, it is more real than that world because of the rigidity of the structure it imposes on World 1. There is no element of chance or contingency in it, and it is certainly not socially constructed (although our understanding of it is so constructed). It rigidly constrains what can happen in the physical world.

If one believes that physical laws are prescriptive, rather than descriptive (cf. the discussion below), one can view the world of all physical possibilities as being equivalent to a complete description of the set of physical laws (for these determine the set of all possible physical behaviors through the complete set of their solutions). The formulation given here is preferable in that it avoids making debatable assumptions about the nature of physical laws, but still incorporates their essential effect on the physical world. Whatever their ontology, what is possible is described by physical laws such as those identified in Table 3.

Abstract (Platonic) Reality

World 4 is the Platonic world of (abstract) realities that are discovered by human investigation, but that are independent of human existence. They are not embodied in physical form, but they can have causal effects in the physical world.

As discussed above, the existence of a world of mathematical objects is strongly argued by Penrose (1997, 2005) and Changeux and Connes (1998). It is a quite different kind of existence from that of physical entities: it is eternal and unchanging, and so is of a Platonic character.

Existence and Epistemology

The major proposal (Ellis 2004) is that *all these worlds exist – Worlds 1 through 4 are ontologically real and are distinct from one another*, as argued above. These claims are justified in terms of the effectiveness of each kind of reality in influencing the physical world.

What then of epistemology (the nature of our knowledge of how things are)? Given the existence of the 4 Worlds as described above, the proposal here is that *epistemology is the study of the relation between World 2 and Worlds 1, 3, and 4*. It attempts to obtain accurate correspondences to quantities in all the worlds by means of entities in World 2.

Table 3

Three Fundamental Laws of Nature

The Second Law of Thermodynamics
Entropy always increases
Maxwell's Laws of Electromagnetism
Electric charges cause electricity and magnetism, which in turn determine the motion of the charges
Einstein's Law of Gravitation
Matter causes spacetime curvature

This exercise implicitly or explicitly divides World 2 theories and statements into (1) true/accurate representations, (2) partially true/misleading representations, (3) false/poor/misleading representations, and (4) representations where we do not know the situation. These assessments range from statements such as 'It is true her hair is red' or 'There is no cow in the room' to 'electrons exist', 'Newtonian theory is a very good description of medium-scale physical systems at low speeds and in weak gravitational fields', and 'The evidence for UFOs is poor'. This raises interesting issues about the relation between reality and appearance: for example, everyday life gives a quite different appearance to reality than microscopic physics – as Eddington (1928) poited out, a table is actually mostly empty space between the atoms that make up its material substance; but in our experience it is a real, hard object. As long as one is aware of this, it can be adequately handled.

4. The Nature of the Laws of Nature

Given this understanding of the kinds of existence, what can we say about the laws of nature? These are impersonal dynamical principles operating on material entities in a spacetime that are needed for reliable emergence of complex orders of existence, such as galaxies, stars, planets, and life, as discussed above. This emergence can happen only if there are precisely ordered evolving relationships and processes governing the behavior of matter at the basic levels that are such as to allow the higher levels to emerge.

The Basic Laws

The hierarchy of complexity is based at the lower levels in specific families of particles (Cottingham and Greenwood 2007), interacting through 4 fundamental forces that are presumed unified at high energies. Their behavior is based in quantum mechanics, with interactions describable by variational principles subject to special relativity theory and to fundamental symmetries entailing conservation laws, but with the exact symmetries of the theory broken. These interactions involve specific masses and interaction strengths. All of this takes place in a 4-dimensional Riemannian spacetime, whose geometry is determined by the matter content of the spacetime.

Emergent Laws

It is far from obvious that these fundamental laws can lead to the emergence of complexity. Regularities of behavior at the higher levels of the hierarchy are characterized by emergent laws at that level; for example, the perfect gas law, Ohm's law, Bernoulli's law, and so on are higher-level physical laws. Each has a domain of validity representing the conditions where it gives reliable results (for example, the temperature and density must be neither too high nor too low for each of these laws to be true). Within their domain of validity, these are equally as valid as the fundamental laws, for they can be thoroughly tested and shown to be accurate predictors of the relevant physical behavior. Similarly, Darwin's theory of evolution through natural selection is an emergent law at the biological population level, as are the laws of genetic inheritance, whereas the Hodgkin–Huxley equations for a nerve's action potential are an effective biological law at the physiological level. The issue of whether there are reliable psychological laws is more uncertain (for a discussion, see Silverberg 2004); but there certainly are universal regularities applicable to human life, such as the need for food and the inevitability of death.

Actually, all the physical laws of which we are certain (such as those in Table 3) are emergent, rather than fundamental laws; for example, both Newton's and Einstein's laws of gravity are presumed to be approximations to some deeper underlying quantum theory of gravity. Indeed, we do not even know what the fundamental physical laws are (string theory/M theory is a popular contender, but it does not yet have a unique formulation and is not experimentally tested).

Table 4

Examples of the Geometrical Nature of the Laws of Physics

Geometric conservation of particles or fields in a 3-dimensional space gives an inverse-square law.
Parallel transport along curves underlies Yang–Mills theory, the basis of the standard model of particle physics.
The Aharanov–Bohm effect and Feynman path integrals suggest that holonomy (based on parallel transport along closed curves) is a fundamental entity in quantum theory.
Geodesic paths (corresponding to extreme distances) underlie variational principles, as a geometrical description of solutions of dynamical equations.

Their Mathematical Description

A mathematically useful description is a probable outcome of reliable behavior, such as is described by physical laws. At a certain level, this is an answer to Wigner's (1960) famous question, *Why are physical laws so well describable by mathematics?*

As mathematics describes ordered patterns of relationships, it is perhaps not surprising that these relationships and processes can be described mathematically. The very nature of mathematics is indeed to describe patterns (Devlin 1996): in space and in time, and indeed within patterns (leading to recursion and higher-order relations). What is surprising is that *fundamental physical relationships can often be described so accurately by very simple laws*, such as an inverse-square law. My suggestion is that *this is because the underlying nature of these fundamental laws is geometrical*, which results in their being accurately represented by simple analytic relations. Some examples are shown in Table 4.

It is a basic principle of mathematics that geometrical relations can be represented in analytic form, and vice versa; the proposal here is that the geometrical form of physical relationships is the more fundamental, and that is the reason for the relatively simple analytic forms for physical laws.

The Nature of the Laws of Physics

Two major issues arise in regard to the existence of the laws of physics. The first is the nature of these laws: are they *descriptive, just characterizing the way things are*, or are they *prescriptive, forcing them to be this way* (Carroll 2004)? For example, quantum field theory applied to the standard model of particle physics is immensely complex (Peskin and Schroeder 1995; see Table 5).

What is the nature of existence of all this quantum apparatus? Are they just descriptions devised by our minds, or are they the way things really are? Derived (effective) theories, including classical (nonquantum) theories of physics, equally have complex abstract structures underlying their use: force laws, interaction potentials, metrics, and so on. The same issue arises.

If laws are descriptive, this is just the way matter behaves; that is, the laws are phenomenological: they just describe what is. They are mathematical and physical constructs that happen to characterize reasonably accurately the physical nature of physical quantities. That such descriptions should be accurate predictors of physical behavior is in itself remarkable. But then from where do the properties of matter derive, leading to these descriptions? How are its characteristic behaviors enforced? The specific issue arising is, *Why does all matter have the same properties wherever it exists in the Universe?* Why are electrons here identical to those at the other side of the Universe if the laws are only descriptive? We seem to have no handle with which to investigate such questions. But we can claim that even if this is their nature, the laws have an existence independent of human minds: the phenomenological descriptions are accurate and will be derived as such by any other intelligent beings in the Universe. In this sense, they are as ontologically real as the laws of mathematics, discussed above: they exist universally as descriptions of the nature of reality.

Table 5

Entities Assumed to Exist When Quantum Field Theory Applies

Hilbert spaces, operators, commutators, symmetry groups, higher--dimensional spaces;
Particles/waves/wave packets, spinors, quantum states/wave functions;
Parallel transport/connections/metrics;
The Dirac equation and interaction potentials, Lagrangians, and Hamiltonians;
Variational principles that seem to be logically and/or causally prior to all the rest.

If they are prescriptive, the laws of physics somehow exist in a form that enables them to control the nature of existence. They then represent a fundamental underlying reality, as entities that have the power to control the behavior of physical quantities. Then matter will necessarily be the same everywhere because their behavior is determined by these laws (assuming that the laws themselves are invariable). The issue arising then, is, *In what way – where and how – do laws of physics exist and impose themselves on the matter in the Universe?* Do they, for example, have an existence in some kind of Platonic space that controls the nature of matter and existence?

The second issue, related to the first, is, *Do these laws in some sense precede the existence of the Universe, somehow governing its coming into being, or do they come into being with the Universe?* This is where this issue relates deeply to the nature of cosmology.

Many theories of the creation of the Universe assume that all these laws, or at least a basic subset of them, preexist the coming into being of the physical Universe because they are presumed to underlie the creation process; for example, the entire apparatus of quantum field theory mentioned above is often taken for granted as preexisting our Universe. This is, of course, an unprovable proposition, but it appears to be widely held. If it is true, it seems to me to support the idea that the laws are indeed ontologically real: indeed, more real than the Universe itself, which is then a transient manifestation of this underlying immutable reality. And it is difficult to see any other basis for the coming into being of matter effectively obeying physical laws.

The Ontological Nature of Effective (Higher-Level) Laws

As discussed in 'Emergent Laws' above, we actually deal with effective laws only: we do not yet have access to the fundamental laws. Hence, my position is that the existential status of fundamental and effective laws has to be the same: they are both genuine representations of the way things behave at the different levels and also preexist the coming into being of any matter that materializes their nature or any mind that comprehends them. In this way, they are ontologically real (i.e., they actually exist), whether descriptive or prescriptive. And this applies at each level of the hierarchy: chemical laws and biological laws are just as real as physical laws, independent of the existence of any human mind.

This ontological reality can be shown by their causal effectiveness. Consider how Maxwell's *theory* of electromagnetism (an abstract entity, described by Maxwell's equations) led to the development of radio, and then to the existence of TV, cell phones, and so on. Maxwell's theory is not the same as any single person's brain state. It can be represented in many ways (on blackboards, in print, on computer screens, in spoken words, in neural connections) and many formalisms (via 3-dimensional vectors or 4-dimensional tensors, for example); these various representations together form an equivalence class, as they all lead to the same predicted outcomes. How do you demonstrate this kind of causation? Design an artifact such as a cell phone through use of Maxwell's theory, and then construct it and operate it. The abstract theory will have altered physical configurations in the real world and hence is causally effective. It is therefore an ontologically existing entity, successfully representing the nature of physical reality in a way that will be accessible to advanced civilizations anywhere in the Universe. The same is true, for example, of the laws of thermodynamics.

As noted in 'Physical and Biological Possibilities' above, one can avoid talking about the laws of behavior *per se* by instead considering the *space of possibilities* underlying what exists physically, rigorously constraining the possible natures of what actually comes into existence. This space is more or less uniquely related to the underlying laws in the same way that the space of solutions of differential equations is related to the nature of the equations. This enables one to avoid the issue of the ontology of the laws, but does not solve it. It does confirm, however, that we can think of the space of possibilities, essentially representing the outcomes, and, hence, nature of the underlying laws, as genuinely existing: it is a transcendent eternal reality, governing the nature of what can actually come into existence.

5. Ultimate Reasons

Philosophers have debated for millennia whether the ultimate nature of existence is purely material or embodies some form of rationality (*logos*) and/or purpose (*telos*). What in the end underlies it all? Is the ultimate nature of the Universe purely material, or does it in some way have an element of the mental? That profound debate is informed by physical cosmology, but cannot be resolved by the physical sciences alone. Given suitable lowest-level laws, with restricted structure and coupling constants (Davies 1982, Barrow 2003), a hierarchy of complexity with effective higher-level laws can emerge. But what essentially underlies the lowest-level laws on which the rest is based? Why do they exist, with the form they have that allows life to exist? Is the ultimate underlying reason pure chance, probability, necessity, or purpose? I consider each of these possibilities in turn.

Pure Chance, Signifying Nothing

The initial conditions in the Universe just happened, and they led to things being the way they are now, by pure happenstance. This is just the way it was: there is no suggestion that it was a probable outcome of some underlying dynamics. Probability does not apply. There is no further level of explanation that applies; searching for 'ultimate causes' has no meaning.

This is a logically possible ultimate reason, but has no further explanatory power; indeed, it is denial that at a fundamental level there is any explanation and so is unsatisfactory to almost everyone (whether scientifically or religiously inclined), primarily because we know that explanations (both impersonal and personal) do indeed exist in the social and mental world. Furthermore, it is difficult to resist the argument that the outcome is so unlikely that pure chance simply is not credible as a reason. Not merely qualia and emotions, but also complex theories such as Einstein's theory of relativity and quantum field theory have come into existence as extraordinarily complex theoretical constructs. To suggest that these can all arise without existence of any underlying cause, or can come into existence out of pure chaos or nothingness without any further guiding structure, seems simply absurd; but if you have such a cause or guiding structure, you do not have pure chaos or nothingness.

This is certainly logically possible and, indeed, is philosophically unassailable; but it is not satisfying as an explanation, as we obtain no unification of ideas or predictive power from this approach. Nevertheless, some implicitly or explicitly hold this view.

High Probability

The idea is that although the structure of the Universe appears to be very improbable, in fact for physical reasons it is highly probable. This is often realized by some version of the idea of *universality*: 'All that is possible, happens'. The current embodiment of this idea is via the concept of a Multiverse: an ensemble of universes or of disjoint expanding universe domains is realized, in which all possibilities occur (Tegmark 1998, 2004, Rees 2000, 2001). It is then supposed that the anthropic principle is realized in both its strong form (if all that is possible happens, then life must happen: it is inevitable) and its weak form (life will occur only in some of the domains that are realized; these are picked out from the others by the fact that we can exist only where life is possible, so the anthropic principle is viewed as a selection principle in the Multiverse context). The favored cosmological setting for this idea is chaotic inflation (Linde 1986, 1990, Vilenkin 2006, Guth 2007, Weinberg 2007), sometimes integrated with the idea of the landscape of string theory (Susskind 2005).

These arguments are only partially successful, even in their own terms. They run into problems if we consider the full set of possibilities: discussions proposing this kind of view actually implicitly or explicitly restrict the considered possibilities *a priori*, for otherwise it is not very likely that the Universe will be as we see it (inflation does not in fact solve the issue of probability; see Penrose 1989a, 2005). Besides, we do not have a proper measure to apply to the set of initial conditions, enabling us to assess these probabilities. Furthermore, the Multiverse hypothesis is not observationally or experimentally provable: the supposed other universe domains are not observable by any conceivable form of astronomical observation, as they lie beyond the visual horizon, and the assumed underlying physics is unproven and probably untestable (Carr and Ellis 2008, Ellis 2009).

Despite these problems, this approach has considerable support in the scientific community, particularly via the chaotic inflationary proposal. It is an attractive explanatory proposal, but is not testable physics; and it is, in fact, not an ultimate explanation. If it were to exist, the whole issue of probability arises again as regards the Multiverse (why this one, rather than another?), leading to the specter of infinite regress: we might explain the probability of a specific Multiverse by assuming an ensemble of multiverses. For the scientist, probability trumps a lack of any explanation. But probability by itself is always an incomplete explanation: for what underlies the laws that govern those probabilities? Why are the assumed laws underlying probabilistic calculations valid in the first place? In any case, probability is a good explanation for intermediate levels of explanation where the dynamical laws have already been established and some kind of variation takes place through either irreducible uncertainty (quantum mechanics) or statistical variation, but there is no evidence that it applies in the context of the Universe and ultimate causation; indeed, the very concept of probability is not applicable if there is only 1 object (the unique Universe) in existence. Application of probability arguments to the Universe itself is dubious because the Universe is unique (Ellis 2006).

In brief: the unique Universe that actually exists may well not be probable, as was taken for granted in the past; indeed, it certainly is improbable because it is of such a nature as to allow life to exist. The whole Multiverse endeavor is an attempt to make the improbable appear probable. It only postpones the problem; it does not solve it.

An interesting variant is the idea of *cosmological natural selection* (Smolin 1992). If a process of re-expansion after collapse to a black hole were properly established, it would open the way to the concept of evolution of the Universe not merely in the sense that its structure and contents develop in time, but in the sense that the Darwinian selection of expanding Universe regions could take place. The idea is that there could be collapse to black holes followed by reexpansion, but with an alteration of the constants of physics through each transition, so that each time that there is an expansion phase, the action of physics is a bit different. The crucial point then is that some values of the constants will lead to production of more black holes, while some will result in less. This allows for evolutionary selection favoring the expanding Universe regions that produce more black holes (because of the favorable values of physical constants operative in those regions), for they will have more 'daughter' expanding Universe regions. Thus, one can envisage natural selection favoring those physical constants that produce the maximum number of black holes.

This is an intriguing effort to bring ideas of natural selection, as discussed in 'The Causal Efficacy of Nonmaterial Entities' above, into cosmology; but again it is only partially successful (Rothman and Ellis 1992) and does not solve the ultimate issues: if this indeed takes place, then why the laws of physics that allow it to take place?

Necessity

This is the idea that things have to be the way they are; there is no other option. The features we see and the laws underlying them are demanded by the unity of the Universe. Coherence and consistency require that things must be the way they are; the apparent alternatives are illusory. Only 1 kind of physics is self-consistent: all logically possible universes must obey the same physics.

To show that this is the case has been one of the most profound goals of theoretical physics, following up powerful unification principles with the aim of eventually devising a single theory of fundamental physics with no free parameter whatever. To really succeed would be a wonderful achievement, potentially leading to a self-consistent and complete scientific view of the foundations of physical existence. But we can imagine alternative universes! – why are they excluded? Indeed, this project seems to fail in any case because fundamental physics is presently going the other way: the hoped-for uniqueness of fundamental theories has evaporated and been replaced by the multiple billions of possibilities of string vacua (as explained by Susskind 2005). Furthermore, here we run into the problem that we have not succeeded in devising a fully self-consistent view of physics: neither the foundations of quantum physics nor those of mathematics are on a really solid consistent basis. Until these issues are resolved, this line cannot be pursued to a successful conclusion.

Additionally, if this approach ever succeeded, it would in fact worsen, rather than solve the fine-tuning problem. If there were just in the end 1 set of constants of nature that are consistent with one another, why should they take those precise improbable values that just happen to allow the existence of life? In the end, this is highly implausible. How could it be that the existence of love and pain and intellect is *necessarily* written into variational principles and mathematical symmetries such as SU(10) or E8 as the inevitable outcome of the only consistent possibility for their implementation?

In any event, just as in the previous case, this is not an ultimate answer: the attempt to implement necessity leaves unexplained the choice of those specific realized features of physics that then lead to the necessity. Why should physics have the specific restricted nature that leads to the necessity of particular high-level features?

This whole project is an attempt to implement the idea of necessity underlying existence and causation, but it necessarily has to be incomplete. What has to be explained includes the following: *Where do the very causal categories of chance, necessity, and purpose come from? How do these concepts arise and have meaning, and what underlying ontological entities or causation do they represent? Why are they themselves necessary? How can they even be relevant, as this whole discussion supposes, if there is no ontological referent that makes the dichotomy between them a meaningful issue?* That itself surely cannot be shown to be necessary: for it is the very category of necessity that has to be explained.

Purpose or Design

The symmetries and delicate balances we observe require an extraordinary coherence of conditions and cooperation of causes and effects, suggesting that in some sense they have been purposefully designed. That is, they may show evidence of intention, both in the setting of the laws of physics and in the choice of boundary conditions for the Universe, in such a way that life will inevitably come into being.

Unlike all the other options just discussed, this introduces an element of meaning, of signifying something, at the foundations. In all the other options, life exists by accident, as a chance byproduct of processes blindly at work. Here, the view is that the totally different quality of existence that emerges in human life from the underlying physics, and the huge fine-tuning that is needed for this to occur, suggests an underlying intention or purpose that this should indeed be the case: *it was meant to be that way*. The modern version, consistent with all the scientific understandings of causation, would see some kind of purpose underlying the existence and specific nature of the laws of physics and the boundary conditions for the Universe, in such a way that life (and eventually humanity) would then come into existence through the operation of those laws, then leading to the development of specific classes of animals through the process of evolution as evidenced in the historical record.

This is the only line of reasoning that does not just relegate the problem to a deeper level: assuming a higher intention is realized through physical reality by design enabling higher-level purpose and ethical principles to be embodied through the nature of the resulting possibility spaces. It is this higher-level set of purposive principles – the underlying *telos* – that is then the ultimate cause of both existence and its specific nature. It is unlikely that this kind of underlying intention could be effective, with emergence of a physical structure where purpose can be deployed so that ethical behavior is meaningful, unless on the one hand the lower-level laws had the kind of impersonal regular behavior that allows reliable higher-level behavior to emerge, thus allowing a mathematical description, and on the other hand something like quantum uncertainty were present so as to free the higher levels from total lower-level determinism. A layered structure emerges: purpose underlies impersonal laws that underlie the emergence of purpose. Two kinds of causation, intentional and impersonal, which undoubtedly both exist in the world around us, occur in an intertwined way, with chance events intervening and helping to lead to the richness of outcomes we see around us.

Thus, implementing this proposal necessarily invokes the other 2: meaningful purpose entails both necessity and chance. Each of these kinds of causation (chance in the sense of probability, necessity, and purpose) does indeed occur in the world in various contexts, but the only one that seems to entail the possibility of being a deep foundation for the others is purpose. There will be some who will reject this possibility out of hand, as meaningless or as unworthy of consideration. However, it is certainly logically possible. Given an acceptance of evolutionary development, it is precisely in the choice and implementation of particular physical laws and initial conditions, allowing such development, that the profound creative activity occurs; and this is where one might conceive of design taking place.

This then relates to religious or spiritual views of the nature of reality, supported by a variety of evidence relative to those domains (that evidence certainly exists, with a wide variety of natures and in a wide variety of contexts; the argument is about its acceptability in each case).

6. The Nature of Evidence

What Kind of Evidence Is Relevant?

In examining these fundamental issues, one needs to take into account all the scientific evidence about the nature of physics, chemistry, and biology that comes from laboratory experiments, as well as about the nature of the Universe that comes from astronomical observations. In particular, I suggest that we are entitled to take the nature of the possibility space allowing existence of material entities supporting consciousness and purposive causation as evidence concerning the ultimate reasons the basic laws are as they are. But equally, as well as the discoveries attained by the scientific method, one should take into account data about the natures of our existence that come from our daily lives and the broad historical experience of humanity (our experiences of ethics and aesthetics, for example).

The claim I make is that the kinds of personal world experience we each have are certainly data on the nature of reality because we live in and indeed are part of reality. They do not have the quality of the strictly repeatable experiments that science engages in – they are much richer than that. As an example, many writings claim that there is no purpose in the Universe: it is all just a conglomerate of particles proceeding at a fundamental level in a purposeless and meaningless algorithmic way. But, I would reply, the very fact that those writers engage in such discourse undermines their own contention; they ignore the evidence provided by their own actions. There is certainly meaning in the Universe to this degree: *the fact that they take the trouble to write such contentions is proof that they consider it meaningful to argue about such issues*; this quality of existence has emerged out of the nature of the physical Universe. Indeed, the human mind is causally effective in the real physical world precisely through many activities motivated by meanings perceived by the human mind.

Any attempt to relate physics and cosmology to ultimate issues must take such real-world experience seriously; otherwise, it will simply be ignoring a large body of undeniable data. These data do not resolve the ultimate issues, but they do indicate dimensions of existence that indeed occur. I consider briefly here 2 types of such evidence that reinforce the arguments given above, namely, concerning experiences related to ethical issues and experiences that many interpret as related to a spiritual or transcendent existence.

The Nature of Ethics

A key area where human experiences are important is ethics and morality: the issue of how we ought to conduct our lives. The origin of moral values has been the subject of debate for centuries. I will just state my own view here and give some references to literature where it is supported. I take the position of moral realism, which argues that we do not invent ethics, but discover it: there is a standard of morality that exists (in 'reality', just waiting to be discovered) that is valid in all times and places, and human moral life is a search to understand and implement that true nature of morality. Thus, I believe that *moral choices relate to ethical values that are timeless and culture independent*, and this is based in existence of a moral reality, as well as a physical reality and a mathematical reality, that underlies the Universe.

For detailed arguments in support of this contention, based in the idea that morality is real and so not just a social construction or evolutionary invention, see Murphy and Ellis (1995), Gaita (2004), and Ellis (2008a). The only point I would make here is the following: there is in fact tacit agreement with this position in the writings of both

Richard Dawkins (2006) and Viktor Stenger (2007), for both make strong claims about the evil caused by religion. In doing so, they are presuming to make a claim that is more than just their personal opinion: they are expressing this as irrefutable fact. Indeed, Stenger claims (p. 216) that it is a scientific fact that evil exists. So what is the experiment that establishes this as a *scientific* fact? There is none, as science does not comprehend the concepts of 'right' and 'wrong'; there are no scientific units ('milli-Hitlers') for degrees of evil. This is an ethical claim falsely dressed up as science. But the point is that this argument by both Dawkins and Stenger does show a belief in absolute standards of right and wrong, independent of culture and space and time, which is in agreement with my own position that underlying the Universe is a moral reality. We can then add to the 4 Worlds listed above in 'The Nature of Existence' another one:

World 5 is the world of Platonic ethical forms, providing a foundation for our sense of ethics. This is perhaps related to a sense of beauty and aesthetics, and it is certainly linked to and based in concepts of purpose and meaning (*telos*), for these are what underlie ethical behavior.

Then the family of worlds we need to recognize becomes as shown in Table 6.

Table 6

The Different Kinds of Reality Enhanced by a Layer Representing Ethical Values and Meaning

World 1:	Matter and Forces
World 2:	Consciousness
World 3:	Physical and Biological Possibilities
World 4:	Abstract (Platonic) Reality
World 5:	Telos – Ethical Reality Related to Purpose

The core of being is then, in this view, the underlying ultimate purpose relating to meaning and morality. One can argue that this purpose, as identified by the spiritual traditions of all the major world religions, is unselfish love, or *agape*. The kind of ethics that is compatible with this view is *kenotic* (self-emptying or purgative), invoking a loving attitude and respect for the freedom and integrity of others as a basic principle underlying the nature of existence, extending to sacrificing on their behalf, a free and willing gift when this has the effect of transforming the situation to a higher ethical level (Ellis 1993, 2008a, Murphy and Ellis 1995).

This quality of existence, whose transforming nature has a self-authenticating quality when experienced, is then seen, in this viewpoint, as the ultimate purpose of the whole, enabled by the nature of the laws of physics and the boundary conditions for the Universe that are set so as to make this possible, and indeed requiring that they be this way in order that these purposes can be realized (Ellis 1993). The core of being, and its ultimate purpose, is then the possibility and the demonstration of caring love for others, as many religious standpoints have proclaimed; this is what leads to and underlies the nature of existence. The material is the vehicle by which this is made possible.

Experiences of the Transcendent

For many, a deeply religious worldview is crucial in understanding our lives and setting values, this worldview being based in our personal life experience, including our experience of a faith tradition and community, religious texts, and inspiring leaders. All of these are data that help us understand our situation and our lives, but their nature and significance are strongly contested by those with other beliefs. My focus here is on what I call 'intimations of transcendence': significant experiences in the lives of many people that reinforce the worldview proposed above. They suggest that there is a copious abundance of being, a plenitude that is more than is necessary, underlying the reality of physical existence. This extra dimension of existence, which we sometimes can glimpse in fleeting ways, is in fact an experience of transcendence. I give a number of examples:

> I say to myself as I watch the niece, who is very beautiful: in her this bread is transmuted into melancholy grace. Into modesty, into a gentleness without words. … Sensing my gaze, she raised her eyes towards mine, and seemed to smile. … A mere breath on the delicate face of the waters, but an affecting vision. … I sense the mysterious presence of the soul that is unique to this place. It fills me with peace, and my mind with the words: 'This is the peace of silent realms'. I have seen the shining light that is born of the wheat. (From *Flight to Arras*, by Antoine de St. Exupery 1969)
>
> One day during my last term at school I walked out alone in the evening and heard the birds singing in the full chorus of song, which can only be heard at that time of year at dawn or sunset. I remember now the shock of surprise with which the sound broke upon my ears. It seemed to me that I had never heard the birds singing before and I wondered whether they sang like this all year round and I had never noticed it. As I walked I came upon some hawthorn trees in full bloom and again I thought that I had never seen such a sight or experienced such sweetness before. I came then to where the sun was setting over the playing fields. A lark rose suddenly from the ground beside the tree where I was standing and poured out its song above my head, and then sank still singing to rest. Everything then grew still as the sunset faded and the veil of dusk began to cover the earth. I remember now the feeling of awe which came over me. I felt inclined to kneel on the ground, as though I had been standing in the presence of an angel; and I hardly dared to look at the face of the sky, because it seemed as though it was but a veil before the face of God. (Bede Griffiths, quoted in Taylor 2007, as quoted in Bellah 2008)
>
> I call to mind that distant moment in [the prison at] Hermanice when on a hot, cloudless summer day, I sat on a pile of rusty iron and gazed into the crown of an enormous tree that stretched, with dignified repose, up and over all the fences, wires, bars, and watchtowers

that separated me from it. As I watched the imperceptible trembling of its leaves against an endless sky, I was overcome by a sensation that is difficult to describe: all at once, I seemed to rise above all the coordinates of my momentary existence in the world into a kind of state outside of time in which all the beautiful things I had ever seen and experienced existed in a total 'copresent'; I felt a sense of reconciliation, indeed of an almost gentle consent to the inevitable course of things as revealed to me now, and this combined with a carefree determination to face what had to be faced. A profound amazement at the sovereignty of Being became a dizzying sensation of tumbling endlessly into the abyss of its mystery; an unbounded joy at being alive, at having been given the chance to live through all I have lived through, and at the fact that everything has a deep and obvious meaning – this joy formed a strange alliance in me with a vague horror at the inapprehensibility and unattainability of everything I was so close to in that moment, standing at the very edge of the 'finite'; I was flooded with a sense of ultimate happiness and harmony with the world and with myself, with that moment, with all the moments I could call up, and with everything invisible that lies behind it and has meaning. I would even say that I was 'struck by love', though I don't know precisely for whom or what. (Vaclav Havel, quoted in Bellah 2008)

Perhaps more wonderful still is the way in which beauty breaks through. It breaks through not only at a few highly organised points, it breaks through almost everywhere. Even the minutest things reveal it as well as do the sublimest things, like the stars. Whatever one sees through the microscope, a bit of mould for example, is charged with beauty. Everything from a dewdrop to Mount Shasta is the bearer of beauty. And yet beauty has no function, no utility. Its value is intrinsic, not extrinsic. It is its own excuse for being. It greases no wheels, it bakes no puddings. It is a gift of sheer grace, a gratuitous largesse. It must imply behind things a Spirit that enjoys beauty for its own sake and that floods the world everywhere with it. Wherever it can break through, it does break through, and our joy in it shows that we are in some sense kindred to the giver and receiver. (Rufus Jones 1920)

> I believe with all my heart and mind that there is a spiritual dimension to all being that cannot be encapsulated in scripture or in creed; an essence that loses its creative force when its communication depends upon the use of words alone. It can I think be readily made manifest through metaphor in poem or story; yet I am deeply aware that even when presented in such a form the truth remains partial. For me, there is a reality that lies beyond our presently misdirected concern for the fruits of economic power. It is only when we acknowledge our deeper inner urge to discover meaning in existence that we begin to harvest the fruits of the spirit. Most of us are at least partially aware of epiphanies that come our way from time to time: the emergence when rounding a corner of a breathtaking panorama of mountain, forest and ocean; the sudden sensibility of a zephyr breeze rustling treetops; the scent of jasmine on a shower of rain. I believe there are illuminations far beyond these: intuitions, insights, divinations that are not shaped by the physical senses; the hand of a friend on one's shoulder in a time of trouble; the sudden recognition of a smile in a passing stranger; above all, the wondrous inspiration of the serendipity, synchronicity, and innate knowing in the fabric of our lives. More often than not gifts such as these, which indelibly inscribe themselves upon our memories, are regarded as gifts of God. (Lewis Watling 2006)

Bernard d'Espagnat, winner of the 2009 Templeton Prize, said in an interview that science has helped him to 'justify his impression of a link between beauty and the divine'[4]. He explained: 'When we hear great classical music or look at very great paintings, they are not just illusions but could be a revelation of something fundamental. I would accept calling it God or divine or godhead but with the restriction that it cannot be conceptualised for the very reason that this ultimate reality is beyond any concept that we can construct'. That is a view I would concur with; it is elaborated in Ellis (2008a).

[4] See: http://www.templeton.org/templeton_report/20090415/; also see http://www.timesonline.co.uk/tol/comment/faith/article5918050.ece.

The Weight of Evidence

What does this all add up to? First, even in order to understand just the material world, it can be claimed that one needs to consider forms of existence other than the material only – for example, a Platonic world of mathematics, on the one hand, and a mental world, on the other, both of which can be claimed to exist and to be causally effective in terms of affecting the material. Our understanding of causation will be incomplete unless we take them into account. The fact that they exist shows that there is more to existence than just material things: as argued above, other kinds of things exist and have real causal powers.

Second, there are also extraordinary qualities of life, including purpose, ethics, aesthetics, and meaning. These certainly exist, but what is the ground of their existence? The possibility space arising out of the fundamental laws of existence must in some sense have these things built into it, for they certainly have come into existence. *And they can have the transcendent quality I have indicated through the quotations above. The fact that such a quality can exist is itself a statement about the nature of reality.* In brief, *ex nihilo nihil fit*: the possibility of meaning and ethics cannot arise out of genuine chaos or out of literally nothing; it has to have been built into the foundations that gave physical existence its structure.

In my view, they represent the ultimate nature of reality because it is extraordinarily unlikely that they could have this nature, this quality of existence, by pure happenstance. Invoking a Multiverse of necessity simply postpones facing the ultimate questions: how can such qualities come into being if they do not represent something about the nature of reality, something that even preexisted the existence of the Universe itself – just like the laws of physics, which are themselves abstract rather than physical entities?

7. The Nature of the Argument

So is the ultimate reason underlying the lowest-level laws on which everything else is based pure chance, probability, necessity, or purpose? They are all logically possible. Neither science nor philosophy can give a certain answer as to which is the deep underlying cause of things: metaphysical uncertainty remains (Ellis 2006). However, if one wants to relate one's understanding to the deeper meaning of personal life, the last option has the most traction. The others in the end provide a more tentative relation to morality and meaning, although experience suggests that these exist.

This view cannot be *proved* to be true, but it is supported by much experience that has considerable persuasive power as a whole. We should take into account data from the whole of life, not just physics or astronomy: we are part of the Universe and live in it. When dealing with ultimate meaning, what is relevant is whatever seems to give ultimate meaning in human life. There is indeed purpose in the Universe (for example, we gathered in Beijing to understand its nature a bit better[5]). Either purpose emerges out of nothing, or it is there from the start as the foundation, then being reflected in life. The latter is indeed a possibility for the nature of ultimate causation and the underlying reason for why the laws of physics exist and are as they are (Ellis 1993, 2008a, Murphy and Ellis 1995). This can provide a satisfying explanation for the full depth of life and human experience, in contrast to reductionist materialist explanations that explicitly or implicitly deny the full depth and reality of this profound experience.

Finally, it should be emphasized that this is not a scientific conclusion, nor is the argument presented one that can be sustained on scientific grounds alone; it is a philosophically based conclusion. The issues considered here (the nature of ultimate causation) are not amenable to scientific resolution, precisely because they go beyond the domain where scientific experiments or observations can give

[5] For the *New Vision 400* conference: http://nv400.uchicago.edu/.

a reliable answer. The argument is thus a philosophical or metaphysical one, based securely on current science, but also taking into account wider philosophical and human issues than cannot be handled by science *per se*. Any attempt to adequately tackle the fundamental issues considered here will necessarily be of this nature. If one wishes to deal purely in terms of scientific argumentation, then the above will be beyond what one will consider as legitimate argument. But if one takes that stand, allowing scientifically rigorous explanation alone, one should also carefully refrain from making any statements about issues of ultimate causation, except if they are identified as purely the personal opinion of whoever is making the statement, and without any scientific standing.

I apologize here for not giving references to all the other books and articles that look at these issues and come to the same kind of conclusions: space simply does not allow this. To pursue things further, one can read the works of Ian Barbour, Arthur Peacocke, John Polkinghorne, John Hick, Keith Ward, Holmes Rolston, Nancey Murphy, Philip Clayton, John Haught, and many others. My views have benefited from them all. As an entrance point to this large literature, see Clayton and Simpson (2008).

References

Al-Khalili J., *Quantum: A Guide for the Perplexed*, Weidendfeld and Nicholson, London 2003[2].

Balashov Y.Y., *Resource letter Ap-1: The anthropic principle*, "American Journal of Physics" 1991, no. 54, p. 1069.

Barrow J.D., *The Constants of Nature*, Pantheon Books, New York 2003.

Barrow J.D., Tipler F.J., *The Anthropic Cosmological Principle*, Oxford University Press, Oxford 1986.

Bellah R.N., *The rules of engagements: Communion in a scientific age*, "Commonweal" 2008, no. (September 12), pp. 15–21.

Berger P.L., *Invitation to Sociology*, Anchor Books, New York.1963.

Bishop C.M., *Neural Networks for Pattern Recognition*, Oxford University Press, Oxford 1999.

Booch G., *Object Oriented Analysis and Design with Applications*, Addison-Wesley, New York 1994.

Campbell D.T., *Downward causation*, [in:] *Studies in the Philosophy of Biology: Reduction and Related Problems*, eds. F. Ayala, T. Dobhzansky, University of California Press, Berkeley 1974.

Campbell N.A., Reece, J.B., *Biology*, Benjamin Cummings, San Francisco 2005.

Carr B., Ellis G.F.R., *Universe or multiverse?*, "Astronomy and Geophysics" 2008, no. 49: pp. 2.29–37.

Carroll J., ed., *Readings on Laws of Nature*, University of Pittsburgh Press, Pittsburgh 2004.

Changeux, J.-P., Connes A., *Conversations on Mind, Matter, and Mathematics*, Princeton University Press, Princeton 1998.

Clayton P., Simpson Z., *The Oxford Handbook of Religion and Science*, Oxford University Press, Oxford 2008.

Cottingham W.N., Greenwood D.A., *An Introduction to the Standard Model of Particle Physics*, Cambridge University Press, Cambridge 2007.

Davies P.C.W., *The Accidental Universe*, Cambridge University Press, Cambridge 1982.

Dawkins R., *The God Delusion*, Bantam Books, New York 2006.

Devlin K., *Mathematics: The Science of Patterns*, Holt, New York 1996.

Durrant A., ed., *Quantum Physics of Matter*, Open University, Milton Keynes 2000.

Eddington A.S., *The Nature of the Physical World*, Cambridge University Press, Cambridge 1928.

Ellis G.F.R., *The theology of the anthropic principle*, [in:] *Quantum Cosmology and the Laws of Nature*, eds. R.J. Russell, N. Murphy, C.J. Isham, Vatican Observatory, Vatican City 1993, pp. 367–406.

Ellis G.F.R., *True complexity and its associated ontology*, [in:] *Science and Ultimate Reality: Quantum Theory, Cosmology and Complexity*, eds. J.D. Barrow, P.C.W. Davies, C.L. Harper, Jr., Cambridge University Press, Cambridge 2004, pp. 607–636.

Ellis G.F.R., *Issues in the philosophy of cosmology*, [in:] *Handbook of the Philosophy of Science: Philosophy of Physics, Part A*, eds. J. Butterfield, J. Earman, North Holland, Amsterdam 2006, pp. 1183–1285.

Ellis G.F.R., *Faith, hope and doubt in times of uncertainty*, [in:] *The James Backhouse Lecture: Australia Yearly Meeting of Quakers*, Friends Book Sales 2008a: sales@quakers.org.au.

Ellis G.F.R., *On the nature of causation in complex systems*, "Transactions of the Royal Society of South Africa" 2008b, no. 63, pp. 69–84. Available at: http://www.mth.uct.ac.za/~ellis/Top-down%20Ellis.pdf.

Ellis G.F.R., *Dark matter and dark energy proposals: maintaining cosmology as a true science?* [in:] *Dark Energy and Dark Matter: Observations, Experiments and Theories*, eds. E. Pecontal, T. Buchert, Ph. Di Stefano, et al., European Astronomical Society Publications Series 2009, no. 36, pp. 325–336.

Feynman R., *QED: The Strange Theory of Light and Matter*, Princeton University Press, Princeton 1985.

Flood R.L., Carson E.R., *Dealing with Complexity: An Introduction to the Theory and Application of Systems Science*, Plenum Press, London 1990.

Gaita R., *Good and Evil: An Absolute Conception*, Routledge, London 2004.

Gilbert S.F., *Developmental Biology*, Sinauer Associates, Sunderland, MA 2006.

Goodstein D.L., *States of Matter*, Dover, New York 1985.

Guth A., *Eternal inflation and its implications*, "Journal of Physics A" 2007, no. 40, pp. 6811–6826.

Harold F.H., *The Way of the Cell: Molecules, Organisms, and the Order of Life*, Oxford University Press, Oxford 2001.

Harrison E.R., *Cosmology*, Cambridge University Press, Cambridge 2000.

Jones R., *Where the beyond breaks through*, "The Friend" 1920, no. 60 (new series), p. 26.

Laidler K.J., Meiser J.H., Sanctuary B.C., *Physical Chemistry*, Houghton Mifflin, Boston 2003.

Linde A.D., *Eternally existing self-reproducing chaotic inflationary universe*, "Physics Letters B" 1986, no. 175, pp. 395–400.

Linde A.D., *Particle Physics and Inflationary Cosmology*, Harwood Academic, Newark 1990.

Murphy N., Ellis G.F.R., *On the Moral Nature of the Universe*, Fortress Press, Minneapolis 1995.

Noble D., *The Music of Life: Biology beyond Genes*, Oxford University Press, Oxford 2007.

Peacocke A.R., *An Introduction to the Physical Chemistry of Biological Organization*, Oxford University Press, Oxford 1989.

Penrose R., *Difficulties with inflationary cosmology*, [in:] ed. E.J. Fergus, *14th Texas Symposium Relativistic Astrophysics*, Proceedings of the New York Academy of Science, New York 1989a.

Penrose R., *The Emperor's New Mind: Concerning Computers, Minds and the Laws of Physics*, Oxford University Press, New York 1989b.

Penrose R., *The Large, the Small, and the Human Mind*, Cambridge University Press, Cambridge 1997.

Penrose R., *The Road to Reality: A Complete Guide to the Laws of the Universe*, Knopf, New York 2005.

Peskin M.E., Schroeder D.V., *An Introduction to Quantum Field Theory*, Perseus Books, Reading, MA 1995.

Polkinghorne J., *Quantum Theory: A Very Short Introduction*, Oxford University Press, Oxford 2002.

Popper K., Eccles J., *The Self and Its Brain: An Argument for Interactionism*, Springer, Berlin 1977.

Rees M.J., *Just Six Numbers: The Deep Forces That Shape the Universe*, Basic Books, New York 2000.

Rees M.J., *Our Cosmic Habitat*, Princeton University Press, Princeton 2001.

Rothman T., Ellis G.F.R., *Smolin's natural selection hypothesis*, "Quarterly Journal of the Royal Astronomical Society" 1992, no. 34, pp. 201–212.

St. Exupery A. de, *Flight to Arras*, Mariner Books, San Diego 1969. (Originally published in French, 1942).

Scott A., *Molecular Machinery: The Principles and Power of Chemistry*, Blackwell, Oxford 1989.

Scott A., *Stairway to the Mind*, Springer, New York 1995.

Seife, J., *Zero: The Biography of a Dangerous Idea*, Penguin, London 2000.

Sellars R.W., *The Philosophy of Physical Realism,* Russell and Russell, New York 1932.

Silk J., *On the Shores of the Unknown: A Short History of the Universe*, Cambridge University Press, Cambridge 2005.

Silverberg A., *Psychological laws*, "Erkenntnis" 2004, no. 58, pp. 275–302.

Simon H.A., *The Sciences of the Artificial*, MIT Press, Cambridge, MA 1982.

Smolin L., *Did the universe evolve?*, "Classical and Quantum Gravity" 1992, no. 9, p. 173–191.

Stenger V.J., *God, the Failed Hypothesis: How Science Shows That God Does Not Exist*, Prometheus Books, Buffalo, NY 2007.

Susskind L., *The Cosmic Landscape: String Theory and the Illusion of Intelligent Design*, Little, Brown, New York 2005.

Taylor C., *A Secular Age*, Harvard University Press, Harvard 2007.

Tegmark M., *Is "the theory of everything" merely the ultimate ensemble theory?*, "Annals of Physics" 1998, no. 270, pp. 1–51.

Tegmark M., *Parallel universes*, [in:] *Science and Ultimate Reality: Quantum Theory, Cosmology and Complexity*, eds. J.D. Barrow, P.C.W. Davies, C.L. Harper, Jr., Cambridge University Press, Cambridge 2004, pp. 459–491.

Vilenkin A., *Many Worlds in One. The Search for Other Universes*, Hill and Wang, New York 2006.

Watling L., *What I believe*, "Southern Africa Quaker News" 2006, no. 220, pp. 47–50.

Weinberg S., *Living in the multiverse*, [in:] *Universe or Multiverse?*, ed. B. Carr, Cambridge University Press, Cambridge 2007. Also see: arXiv:hep-th/0511037.

Wigner E., *The unreasonable effectiveness of mathematics in the natural sciences*, "Communications in Pure and Applied Mathematics" 1960, no. 13 (1) (February). New York: John Wiley & Sons, Inc. Available at: http://www.dartmouth.edu/~matc/MathDrama/reading/Wigner.html.

George F.R. Ellis
University of Cape Town

Top-Down Causation as the Key to the Emergence of Complexity

1. The Theme

A key assumption underlying most present day scientific thought is that causation is bottom up all the way: particle physics underlies nuclear physics, nuclear physics underlies atomic physics, atomic physics underlies chemistry, and so on. Thus all the higher level subjects are at least in principle reducible to particle physics, which is therefore the only fundamental science; as famously claimed by Dirac, chemistry is just an application of quantum physics (Schweber 1993).

However there are many topics that one cannot understand by assuming this one-way flow of causation: for example,

- The flourishing subject of social neuroscience makes clear how social influences act down on individual brain structure (Cacioppo and Berntson 2005, Chiao 2009);
- Studies in vision show that the way we see things is modulated top-down in the brain (Frith 2007, Purves 2010, Kandel 2012);
- Studies in physiology demonstrate that the concept of downward causation is necessary in understanding the heart, where this form of causation can be represented as the influences of initial and boundary conditions on the solutions of differential equations used to represent physiological processes (Noble 2012);
- Epigenetic studies demonstrate that biological development is crucially shaped by the environment (Gilbert and Epel 2009).

- Top-down causation is also prevalent in physics (Ellis 2012a), even though this is not often recognised as such.
- It occurs in particular in digital computers. This is of significance because of a tendency in some sectors of the physics community to represent almost all causation as a computation.

Thus my first theme is that

- *The foundational assumption that all causation is bottom up is wrong in general, and in particular in the case of physics.*

I note here that some writers on this topic prefer to refer to use the terms 'contextual effects' or 'whole-part constraints'. Those are perfectly acceptable, but I will make the case that the stronger term 'top-down causation' is appropriate in many cases. This form of causation does not occur by violating physical laws; on the contrary, it occurs through the laws of physics, by setting constraints on lower level interactions.

My second theme is that

- *Emergence of true complexity is not possible through bottom up effects alone: it requires shaping of lower level interactions by top-down causation, and so a reversal of information flow from bottom-up to top-down* (Walker et al 2012. Walker and Davies 2012).

2. Foundations

Some preliminary remarks will set the foundations for what follows.

1. Causation: The nature of causation is highly contested territory, and I will take a pragmatic view:
***Definition 1*: Causal Effect.** *If making a change in a quantity X results in a reliable demonstrable change in a quantity Y in a given context, then X has a causal effect on Y.*

***Example*:** I press the key labelled 'A' on my computer keyboard; the letter 'A' appears on my computer screen.

Now there are of course a myriad of causal influences on any specific event: a network of causation is always in action. What we usually do is to have some specific context in mind where we keep almost all parameters and variables fixed, allowing just one or two remaining ones to vary; if they reliably cause some change in the dependent variable in that context, we then label them as 'the cause'. This produces a valid local theory of causation, You don't have to solve problems of ultimate causation to understand local physical effects (e.g. heating water causes it to boil).

For example in the case of a digital computer, we have

$$(\textit{The laws of physics, a specific computer structure, software loaded, data}) \rightarrow (\textit{output}) \qquad (1)$$

In a particular run on a specific computer, the laws of physics do not change and the high level software loaded (e.g. Microsoft Word) will be fixed, so the above reduces to

$$(\textit{data}) \underset{\textit{MS-Word}}{\rightarrow \rightarrow \rightarrow} (\textit{output}) \qquad (2)$$

If however we load new high level software (e.g. now we run Photoshop) we will end up with a different relation than (2):

$$(\textit{data}) \underset{\textit{Photoshop}}{\rightarrow \rightarrow \rightarrow} (\textit{output}) \qquad (3)$$

Hence both the data and the software are causes of the output, in the relevant context. The laws of physics are also a cause, but we usually don't bother to mention this: we take them for granted.

Thus we get a useful concept of 'the cause' by taking all except a few for granted.

2. Existence: Given this understanding of causation, it implies a view on ontology (existence) as follows (Ellis 2004): I assume that physical matter (comprised of electrons, protons, etc.) exists. Then the following criterion for existence makes sense:

***Definition 2*: Existence** *If Y is a physical entity made up of ordinary matter, and X is some kind of entity that has a demonstrable causal effect on Y as per Definition 1, then we must acknowledge that X also exists (even if it is not made up of such matter).*

This is clearly a sensible and testable criterion; in the example above, it leads to the conclusion that both the data and the relevant software exist. If we do not adopt this definition, we will have instances of uncaused changes in the world; I presume we wish to avoid that situation.

3. Hierarchy of scales and causation: The basic hierarchy of physical matter is indicated in Table 1. Because it indicates physical scales, it is also an indication of levels of bottom-up causation, hence one can call it a hierarchy of causation. The computer hierarchy and life sciences hierarchy are rather different: those hierarchies are based on causation rather than scale (see Tables 2 and 3 below).

Now a core aspect of this hierarchy of emergent properties is that one needs different vocabularies and language at the different levels of description. The concepts that are useful at one level are simply inapplicable at other levels. The effective equations at the various levels are valid at a specific level but are valid for restricted conditions. They are written in terms of the relevant variables at those levels. These emergent variables can sometimes be obtained by coarse graining of lower level states, but not always so.

The key feature underlying this hierarchy is that the higher level dynamics is effectively decoupled from lower level laws and details of the lower level variables (Schweber 1993): with a few exceptions related to structured systems, you don't have to know those details in order to predict higher level behaviour. This is why it even makes sense to talk about the different levels shown there.

Table 1

The hierarchy of physical matter[1].

	Domain	Scale	Mass	Example
L17	Cosmology	10^{26} m	10^{53} kg	Observable Universe
L16	Large Scale Structures	10^{23} m	10^{47} kg	Great Attractor, Sloan Great wall
L15	Galaxy Clusters	10^{22} m	10^{45} kg	Virgo cluster, Coma cluster
L14	Galaxies	10^{21} m	10^{42} kg	M31, NGC 1300, M87
L13	Star clusters	10^{20} m	10^{35} kg	Messier 92, Messier 69
L12	Stellar systems	10^{12} m	10^{30} kg	Binaries, Solar system
L11	Stars	10^{10} m	10^{30} kg	Sun, Proxima Centauri, Eta Carinae
L10	Planets	10^{9} m	10^{24} kg	Earth, Mars, Jupiter
L9	Continents	10^{7} m	10^{17} kg	Africa, Australia
L8	Land forms	10^{4} m	10^{8} kg	Atlas mountains, Andes
L7	Macro objects	1 m	10 kg	Rocks, chairs, computers
L6	Materials	10^{-2} m	10^{-1} kg	Conductors, Insulators, semi-conductors
L5	Molecules/ chemistry	10^{-9} m	10^{-25} kg	H_2O, SiO_2, $C_6H_{12}O_6$, $C_9H_{13}N_5O_{12}P_3$
L4	Atomic physics	10^{-10} m	10^{-26} kg	Hydrogen atom, Carbon atom
L3	Nuclear physics	10^{-14} m	10^{-27} kg	Neutron, Proton, Carbon nucleus
L2	Particle physics	10^{-15} m	10^{-33} kg	Quarks, electrons, gluons
L1	Quantum gravity	10^{-35} m		Superstrings

A sensible view is that the entities at each classical level of the hierarchy (Table 1) are real. A chair is a real chair even though it is made of atoms, which in turn are real atoms even though they are made of a nucleus and electrons, and so on; and you too are real (else you could not read this paper), as is the computer on which you are reading it. Issues of ontology may be unclear at the quantum level, but they are clear at the macro level.

[1] A fuller description is given at http://www.mth.uct.ac.za/~ellis/cos0.html.

One of the characteristics of truly complex systems is that higher level variables are not always just coarse grainings of lower level variables; they are sometimes crucially related to the details of the structure. Hence in highly ordered structures, sometimes changes in some single micro state can have major deterministic outcomes at the macro level (which is of course the environment for the micro level); this cannot occur in systems without complex structure. Examples are:

- a single error in microprogramming in a computer can bring the whole thing to a grinding halt (got that at the moment in my laptop);
- a single swap of bases in a gene can lead to a change in DNA that results in predictable disease;
- a single small poison pill can debilitate or kill an animal, as can damage to some very specific micro areas in the brain.

This important relation between micro structure and macro function is in contrast to statistical systems, where micro changes have no effect at the macro level, and chaotic systems, where a micro change can indeed lead to a macro change, but it's unpredictable.

One might ask 'where is the top?' The answer depends on context: in the case of structure of the human brain, no top is currently identifiable: things near the top seem to be non-localised. In the case of cosmology, one ends up with philosophy because the largest physical scales are unobservable (there are observational horizons in the real universe) so you can say anything you want about it and no on will ever be able to make observations that contradict what you say. So the claims I make are local claims (for any pairs of related levels) and independent of any global claims about any topmost level.

This applies also to the bottom – there may or may not be a bottom-most level; if there is one we don't know what it is (it may be string/M theory, but then again it may not). Indeed all the levels we deal with in ordinary physics are effective levels, not fundamental;

and this does not matter. This is just as well, else we could not do physics.

3. Contextual Effects in Physics, Astronomy, and Cosmology

Contextual effects occur commonly in classical physics. An example is nucleosynthesis in the early universe: macro-level variables (average cosmic densities) determine the expansion rate of the cosmos, which determines the temperature-time relation T(t) that in turn determines the rates of nuclear reactions (micro-level processes) and hence the outcome of nucleosynthesis, leading to macro variables such as the overall mass fraction of Helium and Lithium in the early Universe. Occurrence of this top-down effect is why we can use element abundance observations to constrain cosmological parameters (Dodelson 2003).

More generally, top-down causation happens wherever boundary conditions and initial conditions determine the results. Environmental variables (a macro scale concept) act down to determine the values of physical fields locally:

$$(\textit{Laws of physics, matter/field description, boundary/initial conditions}) \rightarrow (\textit{output}) \tag{4}$$

Hence for specific matter/fields we find

$$(\textit{boundary/initial conditions}) \underset{\textit{matter/field properties}}{\rightarrow \rightarrow \rightarrow} (\textit{output}) \tag{5}$$

Examples are standing waves; hearing the shape of a drum; and the outcomes of the reaction diffusion equation. Physicists often try to minimise such boundary effects through the idealisation of an isolated system; but no real system is isolated either in space or in time. There is nothing new in all this: it's just that we don't usually talk about this as top-down effects. It may be helpful to do so (Noble 2012).

1. Astronomy: In the context of astronomy/astrophysics, there is a growing literature on contextual effects such as suppression of star formation by powerful active galactic nuclei (Page et al 2012). This is a top-down effect from galactic scale to stellar scale and thence to the scale of nuclear reactions. Such effects are often characterised as feedback effects.

2. Cosmology: In cosmology, there are two venerable applications of the idea of top-down effects: Mach's Principle, and the Arrow of Time (Ellis 2002). In each case it has been strongly suggested that boundary conditions on the Universe as a whole are the basic cause of crucial local effects (the origin of inertia, and the local direction of time that characterizes increasing entropy).

The former is not now much discussed, but the latter has strong recent support (Carroll 2010, Penrose 2011). Some will claim this is 'nothing but' the combined effect of all the particles and fields in the very early universe. Well yes and no: the effect is indeed due to all those particles, but it depends crucially on the *specific relationship* between them. Change that relationship from smooth to very clumpy: exactly the same particles will be there, but will have a very different effect. It's the spatial distribution that matters: the relationship between them is the specific cause of the local physical effect of existence of a unique arrow of time. You cannot describe that higher level relationship in terms of the variables and scales relevant at the lower levels in Table 1. And the outcome is crucially important: life would not be possible without a well-established local arrow of time.

A recent example is the idea of chameleon fields: scalar particles with a non-linear self-interaction so that their effective mass depends on the environment (Khoury and Weltman 2004, 2004a). They are possible candidates for both dark energy and dark matter, because their properties are quite different in different contexts.

3. Quantum Physics: Top down effects occur in the context of quantum physics too (Ellis 2012). Here are some examples:

Band Structure: The periodic crystal structure in a metal leads (via Bloch's theorem) to lattice waves, and an electronic band structure depending on the particular solid involved, resulting in all the associated phenomena resulting from the band structure (Ziman 1979). The entire machinery for describing the lattice periodicity refers to a scale much larger than that of the electron, and hence is not describable in terms appropriate to that scale. Thus these effects all exist because of the macro level properties of the solid – the crystal structure – and hence represent top-down causation from that structure to the electron states. This can lead to existence of quasiparticles such as phonons that result from vibrations of the lattice structure; it also leads to Cooper pairs and hence to superconductivity. Because these are all based in top-down action, they are emergent phenomena in the sense that they simply would not exist if the macro-structure did not exist, and hence cannot be understood by a purely bottom-up analysis, as emphasized strongly by Laughlin (1999).

Quantum cooperative effects occur in superconductivity, superfluidity, and the quantum Hall effect. In superconductivity, the electrons – despite their repulsion for each other – form pairs ('Cooper pairs') which are the basic entities of the superconducting state. This happens by a cooperative process: the negatively charged electrons cause distortions of the lattice of positive ions in which they move, and the real attraction occurs between these distortions (Goodstein 1985). The Nobel lecture by Laughlin (1999) comments on the consequent irreducibility of emergence:

> One of my favourite times in the academic year occurs in early spring when I give my class of extremely bright graduate students, who have mastered quantum mechanics but are otherwise unsuspecting and innocent, a take-home exam in which they are asked to deduce superfluidity from first principles. There is no doubt a very special place in hell being reserved for me at this very moment for this mean trick, for the task is impossible. Superfluidity, like the fractional Hall effect, is an emergent phenomenon – a low-energy collective effect of huge numbers of particles that cannot be deduced from the microscopic equations of motion in a rigorous way, and that disappears completely when the system is taken apart (Anderson 1972). … The students feel betrayed and hurt by this experience because they have been trained to think in reductionist terms and thus to believe that everything that is not amenable to such thinking is unimportant. But nature is much more heartless than I am, and those students who stay in physics long enough to seriously confront the experimental record eventually come to understand that the reductionist idea is wrong a great deal of the time, and perhaps always... The world is full of things for which one's understanding, i.e. one's ability to predict what will happen in an experiment, is degraded by taking the system apart, including most delightfully the standard model of elementary particles itself.

The lecture continues to explain how one of the things emergent phenomena can do is create new physics, and particularly the quantum Hall effect (whereby cooperative phenomena create quasi-particles with fractional charge from interactions between particles with integral charge). What Laughlin did not point out is that top-down effects are key to what he describes. They create the possibility of existence of Cooper pairs and so superconductivity.

Caldeira-Leggett model: The Caldeira-Leggett model is a model for a system plus heat reservoir, used for the description of dissipation phenomena in solid state physics (Breuer and Petruccione 2006).

Here the Lagrangian of the composite system **T** consisting of the system **S** of interest and a heat reservoir **B** takes the form

$$L_T = L_S + L_B + L_I + L_{CT}, \tag{6}$$

where L_S is the Lagrangian for the system of interest, L_B that for the reservoir (a set of non-interacting harmonic oscillators), and L_I that for the interaction between them. The last term L_{CT} is a 'counter term', introduced to cancel an extra harmonic contribution that would come from the coupling to the environmental oscillators. This term represents a top-down effect from the environment to the system, because L_I completely represents the lower-level interactions between the system and the environment. The effect of the heat bath is more than the sum of its parts when $L_{CT} \neq 0$, because the summed effect of the parts is given by L_I.

State vector preparation: State vector preparation is key to experimental set-ups, and is a non-unitary process because it can produce particles in a specific eigenstate from a stream of particles that are not in such a state. Indeed it acts rather like state vector reduction, being a transition that maps a mixed state to a pure state.

How can this non-unitary process happen in a way compatible with standard unitary quantum dynamics? The crucial feature is pointed out by Isham (1995): the outcome states are drawn from some collection E_i of initial states by being selected by some suitable apparatus, being chosen to have some specific spin state in the Stern-Gerlach experiment; the other states are discarded. This happens in two basic ways: separation and selection (as in the Stern Gerlach experiment), which is unitary up to the moment of selection when it is not, or selective absorption (as in the case of wire polarisers), which continuously absorbs energy. This top-down effect from the apparatus to the particles causes an effective non-unitary dynamics at the lower levels, which therefore cannot be described by the Schrodinger or Dirac equations.

In such situations, selection takes place from a (statistical) ensemble of initial states according to some higher level selection criterion, which is a form of top-down causation from the apparatus to the particles. This is a generic way one can create order out of a disordered set of states, and so generate useful information by throwing away what is meaningless (Ellis 2012). The apparatus is specifically designed to have this non-unitary effect on the lower level. This example is important in its own right, but also because it points the way to considering similar top-down effects in photon detectors (Ellis 2012). This may be a crucial part of why actual measurements in a realistic context are non-unitary in character.

4. Complex Structures: Digital Computers

Digital Computers Structured systems such as a computer constrain lower level interactions, and thereby paradoxically create new possibilities of complex behaviour. For example, the specific connections between p-type and n-type transistors can be used to create NOR, NAND, and NOT gates (Mano and Kime 2008); these can then be used to build up adders, decoders, flip-flops, and so on. It is the specific connections between them that channels causation and so enable the lower level entities to embody such logic; the physical structure constrains the movement of electrons so as to form a structured interaction network. The key physical element is that the structure breaks symmetry (cf: Anderson 1972), thereby enabling far more complex behaviour than is possible in isotropic structures such as a plasma, where electrons can go equally in any direction.

However hardware is only causally effective because of the software which animates it: by itself hardware can do nothing. Both hardware and software are hierarchically structured (Tanenbaum 1990), with the higher level logic driving the lower level events. The software hierarchy for digital computers is shown in Table 2.

Table 2

The causal hierarchy in a digital computer system (Tanenbaum 1990)

Level 7:	Applications programs
Level 6:	Problem oriented language level
Level 5:	Assembly language level
Level 4:	Operating system machine level
Level 3:	Instruction set architecture level
Level 2:	Microarchitecture level
Level 1:	Digital logic level
Level 0:	Device level

Entering data by key strokes is a macro activity, altering macro variables. This acts down (effect T1) to set in motion a great number of electrons at the micro physics level, which (effect D1) travel through transistors arranged as logic gates at the materials level; finally (effect B1) these cause specific patterns of light on a computer screen at the macro level, readable as text. Thus we have a chain of top-down action T1 followed by lower level dynamical processes D1, followed by bottom up action B1, these actions composed together resulting in a same level effective macro action D2:

$$D2 = B1 \circ D1 \circ T1 \tag{7}$$

This is how effective same-level dynamics D2 at the higher level emerges from the underlying lower level dynamics D1. This dynamics is particularly clear in the case of computer networking, where the sender and receiver are far apart. At the sender, causation flows downwards from the Application layer through the Transport, Network and Link layers to the Physical layer; that level transports binary code to the other computer through cable or wireless links; and then causation flows up the same set of layers at the receiver end, resulting in effective same-level transport of the desired message from source to destination. The same level effective action is not fictitious: it is a reliable determinable dynamic effect relating variables at the macro level of description at the sender and receiver. If this were not the case you would not be able to read this article, which you obtained by such a process over the internet. There was nothing fictitious about that action: it really happened. Emergent layers of causation are real (Anderson 1972, Schweber 1993, Noble 2012).

The result is that the user sees a same-level interaction take place, and is unaware of all the lower levels that made this possible. This is information hiding, which is crucial to all modular hierarchical systems (Booch 1993). The specific flow of electrons through physical gates at the physical level is determined by whether the high level software is a music playing program, word processor, spreadsheet, or whatever: a classic case of top-down causation (the lower level interactions would be different if different software were loaded, cf. (2) and (3) above). Hence what in fact shapes the flow of electrons at the gate level is the logic of the algorithms implemented by the top level computer program (Lafore 2002, MacCormick 2012).

Four crucial points emerge.

A: Causal Efficacy of Non Physical entities: Both the program and the data are non-physical entities, indeed so is all software. A program is not a physical thing you can point to, but by Definition 2 it certainly exists. You can point to a CD or flashdrive where it is stored, but that is not the thing in itself: it is a medium in which it is stored. The program itself is an abstract entity, shaped by abstract logic. Is the software 'nothing but' its realisation through a specific set of stored electronic states in the computer memory banks? No it is not because it is the *precise pattern* in those states that matters: a higher level relation that is not apparent at the scale of the electrons themselves. It's a relational thing (and if you get the relations between the symbols wrong, so you have a syntax error, it will all come to a grinding halt). Furthermore it is not the same as any specific realisation of that pattern. A story can told (and so represented in sound mediated by air vibrations), printed in a book, displayed on a computer screen, attended to in ones mind, stored in one's memories; it is not the same as any of these particular representations, it in itself is an abstract thing that can be realised in any of these ways; and it is the same with computer programs: they are abstract entities that can be physically realised in many different ways, bit no particular one of them is the same as the story itself. They are its varied representations. This abstract nature of software is realised in the concept of *virtual machines*, which occur at every level in the computer hierarchy except the bottom one (Tanenbaum 1990). But this tower of virtual machines causes physical effects in the real world, for example when a computer controls a robot in an assembly line to create physical artefacts.

B: Logical relations rule at the higher levels: The dynamics at all levels is driven by the logic of the algorithms employed in the high level programs (Lafore 2002). They decide what computations take place, and they have the power to change the world (MacCormick 2012) This abstract logic cannot be deduced from the laws of physics: they operate in a completely different realm. Furthermore the relevant higher level variables in those algorithms cannot be obtained by coarse graining any lower level physical states. They are not coarse-grained or emergent variables: they are assigned variables, with specific abstract properties that then mediate their behaviour.

C: Underlying physics allows arbitrary programs and data: Digital computers are universal computers. The underlying physics does not constrain the logic or type of computation possible, which Turing has shown is universal (Copeland 2004). Physics does not constrain the data used, nor what can be computed (although it does constrain the speed at which this can be done). It enables the higher level actions rather than constraining them. The program logic dictates the course of things.

D: Multiple realisability at lower levels. The same high level logic can be implemented in many different ways: electronic (transistors), electrical (relays), hydraulic (valves), biological (interacting molecules) for example. The logic of the program can be realised by any of these underlying physical entities, which clearly demonstrates that it is not the lower level physics that is driving the causation. This *multiple realisability* is a key feature characterising top-down action (Auletta et al 2008). when some high level logic is driving causation at lower levels, it does not matter how that logic is physically instantiated: it can be realised in many different ways. Thus the top-down map T1 in (7) is not unique: it can be realised both in different physical systems, and in different micro states of the same system.

Equivalence classes: This means that we can consider as the essential variables in the hierarchy, the *equivalence classes of lower level states* that all that correspond to the same high level state (Auletta et al 2008). When you control a top level variable, it may be implemented by any of the lower level states that correspond to the chosen high level state; which one occurs is immaterial, the high level dynamics is the same. You even can replace the lower level elements by others, the higher entity remains the same (the cells in your body are now all different than they were 7 years ago; you are made up of different matter, but you are still the same person).

In digital computers, there are such equivalences all over the place:

- at the circuit level: one can use Boolean algebra to find equivalent circuits;
- at the implementation level: one can compile or interpret (giving completely different lower level functioning for same higher level outcome);
- at the hardware level, one can run the same high level software on different microprocessors;
- even more striking is the equivalence of hardware and software in much computing (there is a completely different nature of lower level entities for the same higher level outcomes).

In each case this indicates top-down effects are in action: the higher level function drives the lower level interactions, and does not care how it is implemented (information hiding is taking place).

Hence although they are the ultimate in algorithmic causation as characterized so precisely by Turing (Copeland 2004), digital computers embody and demonstrate the causal efficacy of non-physical entities. The physics allows this, it does not control what takes place. Computers exemplify the emergence of new kinds of causation out of the underlying physics, not implied by physics but rather by the logic of higher level possibilities. This leads to a different phenomenology at each of the levels of Table 2, described by effective laws for that level, and an appropriate language. A combination of bottom up causation and contextual affects (top-down influences) enables their complex functioning.

5. Complex Structures: Life and the Brain

Life: living systems are highly structured modular hierarchical systems, and there are many similarities to the digital computer case, even though they are not digital computers. The lower level interactions are constrained by network connections, thereby creating possibilities of truly complex behaviour. The basic hierarchy is show in Table 2.

Table 3

The causal hierarchy in life: the human sciences (Campbell and Reece 2005)

Level 8:	Sciology/Economics/Politics
Level 6:	Psychology
Level 5:	Physiology
Level 4:	Cell biology
Level 3:	Biochemistry
Level 2:	Chemistry
Level 1:	Atomic physics
Level 0:	Particle physics

Three new key things come in, as compared to the natural sciences (Table 1).

1. Feedback control systems and homeostasis: Homeostatic systems are ubiquitous in biology, for example at the level of cells, in physiology, and in the functioning of the mind. The dynamics are controlled by the goals embodied in feedback control systems rather than by initial conditions (Ellis 2008, 2012). The goals are abstract entities; like computer algorithms, they are realised in physical ways as patterns of interactions or excitations, but are not themselves physical entities. The highest level of goals relate to purpose and meaning, introducing the causal effectiveness of ethics and aesthetics into human life: these set the conditions governing what lower level goals are acceptable or not, and hence govern outcomes in a crucial way (Murphy and Ellis 2005).

2. Adaptive selection is one of the most important types of top-down causation, underlying evolution, development, and functioning in biology (Ellis 2008, 2012). Adaptive selection is top-down because the selection criteria are at a different level than the objects being selected: in causal terms, they represent a higher level of causation. Darwinian selection is the special case when one has repeated adaptive selection with heredity and variation. This is top-down because the result is crucially shaped by the environment [as demonstrated by numerous experiments: e.g. a polar bear is white because the polar environment is white]. However adaptive selection occurs far more widely than that; e.g. it occurs in biological development as well as function (Campbell and Reece 2005, Gilbert and Epel 2009) and even in physics: state vector preparation is an example, as is Maxwell's demon (Ellis 2012a).

The key point about adaptive selection (whether once off or repeated) is that it lets us locally go against the flow of entropy, and this lets us build up useful information. This is why top down causation is crucial to building up real complexity.

3. Non-Physical causation: as in the case of computers, non-physical entities have causal powers in the life sciences hierarchy. At the lower levels, information is causally effective, for example famously in the case of the genetic code. At the higher levels, as just intimated, a hierarchy of goals determine what happens, at the top level being driven by considerations of purpose and meaning that relate in crucial ways to the society in which the individual is imbedded (Berger and Luckmann 1967). Every time a muscle moves, a conscious or unconscious goal determines what happens.

The Brain: Top-down causation is prevalent at all levels in the brain: for example it is crucial to vision (Frith 2007, Purves 2010, Kandel 2012) as well as the relation of the individual brain to society (Chiao 2009), and of course the way the brain controls muscles and limbs according to chosen goals. The hardware (the brain) can do nothing without the excitations that animate it (which we call the mind): indeed this is the difference between life and death. The mind is not a physical entity, but it certainly is causally effective: proof is the existence of the computer on which you are reading this text. It could not exist if it had not been designed and manufactured according to someone's plans, thereby proving the causal efficacy of thoughts, which like computer programs and data are not physical entities. This is a dominant form of causation in the everyday world.

The topmost level in the human motivational system is purpose or meaning. The David Deutsch has a classic comment on the topic in his book *The Fabric of Reality* (Deutsch 1998) as follows[2]:

> Consider one particular copper atom at the tip of the nose of the statue of Sir Winston Churchill that stands in Parliament Square in London. Let me try to explain why that copper atom is there. It is because Churchill served as prime minister in the House of Commons nearby; and because his ideas and leadership contributed to the Allied victory in the Second World War; and because it is customary to honor such people by putting up statues of them; and because bronze, a traditional material for such statues, contains copper, and so on. Thus we

[2] I thank J.C.N. Smith for this quote.

> explain a low-level physical observation – the presence of a copper atom at a particular location – through extremely high-level theories about emergent phenomena such as ideas, leadership, war and tradition.
>
> There is no reason why there should exist, even in principle, any lower-level explanation of the presence of that copper atom than the one I have just given."

Of course much of the mind's operation is unconscious (Kandel 2012) and much is based in emotions (Damasio 2000): these aspects both involve enormously complex interactions between bottom and top-down causation.

6. Room at the Bottom

Given this evidence for top-down causation, the physicist asks, how can there be room at the bottom for top-down causation to take place? Isn't there over-determination because the lower level physics interactions already determine what will happen from the initial conditions?

There are various ways that top down causation can be effective, based in lower level physical operations:

a: Setting Constraints on lower level interactions, which break symmetries (Anderson 1972) and so for example create more general possibilities than are available to cellular automata. This has been explained above in the context of digital computers. It is also the case where effective potentials occur due to local matter structuring (Ellis 2012). These are crucially dependent on the nature of the higher level structure (e.g. crystal symmetries and gate structure) that cannot be described in terms of lower level variables. It is the higher level patterns that are the essential causal variable in solid state physics by creating a specific band structure in solids (hence for example the search for materials that will permit high temperature superconductivity). Which specific lower level entities create them is irrelevant: you can move around specific protons and electrons while leaving the band structure unchanged. This multiple realisability is always the case with effective potentials.

b: Changing the nature of the constituent entities: the billiard ball model of unchanging lower level entities underlying higher level structure is wrong. Hydrogen in a water molecule has completely different properties than when free; electrons bound in atom interact with radiation quite differently than when free; neutrons bound in nucleus have a half life of billions of years but they decay in 11 ½ minutes when free. Assuming that the nature of an entity is characterised by the way it interacts with others, in each case the higher level context has changed the nature of the underlying entities (an effect that is commonplace in biology).

c: Creating constituent entities: in many cases the lower level entities would not exist without the higher level structure. For example, the very possibility of existence of phonons is a result of the physical structure of specific materials (Laughlin 1999). This structure is at a higher level of description than that of electrons. Emergence of higher level entities has clearly occurred when lower level entities cannot exist outside their higher level context (again, a common effect in biology, where symbiosis is rife).

d: Deleting lower level entities. There is not a fixed set of lower level entities when selection creates order out of disorder by deleting unwanted lower level entities or states: top-down action selects what the lower elements will be. This selective top-down process is what underlies state vector preparation in quantum physics (Ellis 2012). It is crucial in evolutionary biology (Gilbert and Epel 2009) and in the mind (Frith 2007, Kandel 2012).

e: Statistical fluctuations and quantum uncertainty: Lower level physics is not determinate: random fluctuations and quantum indeterminism are both in evidence. What happens is not in the throes of iron determinism: random events take place at the micro level. In complex systems, this unpredictable variability can result in an ensemble of lower level states from which a preferred outcome is selected according to higher level selection criteria. Thus top-down selection leading to increased complexity (Ellis 2012) is enabled by the randomness of lower level processes.

7. The Big Picture: the Nature of Causation

The view put here is that the higher levels are causally real because they have demonstrable causal powers over the lower levels. Indeed the issue of ontology is clear at all higher levels: it is only at the lowest levels (where quantum effects dominate) that it gets murky.

I suggest top-down effects from these levels is the key to the rise of genuine complexity (such as computers and human beings):

Hypothesis: *bottom up emergence by itself is strictly limited in terms of the complexity it can give rise to.* ***Emergence of genuine complexity is characterised by a reversal of information flow from bottom up to top down*** (Walker et al, 2012; Walker and Davies 2012).

The degree of complexity that can arise by bottom-up causation alone is strictly limited. Sand piles, the game of life, bird flocks, or any dynamics governed by a local rule (Arrighi and Grattange 2012) do not compare in complexity with a single cell or an animal body. The same is true in physics: spontaneously broken symmetry is powerful (Anderson 1972) but not as powerful as symmetry breaking that is guided top-down to create ordered structures (such as brains and computers). Some kind of coordination of effects is needed for such complexity to emerge. Some physical processes are not emergent but are entailed in a top-down way. For example there is no bottom up process by which the computer memory states embodying a Quicksort algorithm can emerge from the action of the underlying physics acting in a purely bottom-up way.

The assumption that causation is bottom up only is wrong in biology, in computers, and even in many cases in physics, for example state vector preparation, where top-down constraints allow non-unitary behaviour at the lower levels. It may well play a key role in the quantum measurement problem (the dual of state vector preparation) (Ellis 2012). One can bear in mind here that wherever equivalence classes of entities play a key role, such as in Crutchfield's computational mechanics (Crutchfield 2011), this is an indication that top-down causation is at play.

There are some great discussions of the nature of emergent phenomena in physics (Anderson 1972, Schweber 1993, Laughlin 1999, Hartmann 2001), but none of them specifically mention the issue of top down causation. This paper proposes that recognising this feature will make it easier to comprehend the physical effects underlying emergence of genuine complexity, and may lead to useful new developments, particularly to do with the foundational nature of quantum theory. It is a key missing element in current physics.

What are the major counter arguments? Firstly physicists recognise that all matter is controlled at the bottom level by the forces between particles; hence physics underlies all. There seems to be no room for any other kind of causation. I respond to that claim in Section 6 above: essentially the context determines how the fundamental interactions work out; they offer opportunities and constraints but do not by themselves determine the outcome.

Secondly, this bottom-up view is then taken as an underlying principle of faith by hard core reductionists, who discount the contextual effects that in fact occur: for example claiming that biology is controlled bottom up by genes alone, thereby ignoring all the discoveries of epigenetics, which prove this false. But such reductionism is always a cheat, because it is always only partial. An example: Francis Crick famously wrote

> You, your joys and your sorrows, your memories and your ambitions, your sense of personal identity and free will, are in fact no more than the behavior of a vast assembly of nerve cells and their associated molecules.

But nerve cells and molecules are made of electrons plus protons and neutrons, which are themselves made of quarks … so why not

> You, your joys and your sorrows, your memories and your ambitions, your sense of personal identity and free will, are in fact no more than the behavior of a vast assembly of quarks and electrons?

And these themselves are possibly vibrations of superstrings. Why does he stop where he does? – because that's the causal level he understands best! – he's not a particle physicist. If he assumes that the level of cells and molecules is real, it's an arbitrary assumption unless *all* levels are real – which is my position. It's the only one that makes sense.

In his book *The Character of Physical Law* (1960), on pp. 124–125, Richard Feynman summarises the hierarchy of structure, starting with the fundamental laws of physics and their application to protons, neutrons, and electrons, going on to atoms and heat, and including waves, storms, stars, as well as frogs and concepts like 'man', 'history', 'political expediency', 'evil', 'beauty', and 'hope'. He then says the following (pp. 125–126):

> Which end is nearer to God, if I may use a religious metaphor. Beauty and hope, or the fundamental laws? I think that the right way, of course, is to say that what we have to look at is the whole structural interconnection of the thing; and that all the sciences, and not just the sciences but all the efforts of intellectual kinds, are an endeavour to see the connections of the hierarchies, to connect beauty to history, to connect history to man's psychology, man's psychology to the working of the brain, the brain to the neural impulse, the neural impulse to

> chemistry, and so forth, up and down, both ways. And today we cannot, and it is no use making believe we can, draw carefully a line all the way from one end of this thing to the other, because we have only just begun to see that there is this relative hierarchy.
>
> And I do not think either end is nearer to God. To stand at either end, and to walk off that end of the pier only, hoping that out in that direction is the complete understanding, is a mistake. And to stand with evil and beauty and hope, or with fundamental laws, hoping that way to get a deep understanding of the whole world, with that aspect alone, is a mistake. It is not sensible for the ones who specialize at one end, and the ones who specialize at the other, to have such disregard for each other ... The great mass of workers in between, connecting one step to another, are improving all the time our understanding of the world, both from working at the ends and from working in the middle, and in that way we are gradually understanding this tremendous world of interconnecting hierarchies.

This chapter suggests that it explicit recognition of the intermingling of bottom-up and top-down causation will be of help in this enterprise of understanding these interconnections.

References

Anderson P.W., *More is different*, "Science" 1972, no. 177, pp. 393–396.
Arrighi P., Grattange J., *The quantum game of life*, "Physics World" 2012, no. June, pp. 23–26.
Auletta G., Ellis G.F.R., Jaeger L., *Top-down causation by information control: from a philosophical problem to a scientific research program*, "J. R. Soc. Interface" 2008, no. 5, pp. 1159–1172.
Booch, G., *Object-Oriented Analysis and Design with Applications*, Addison-Wesley Professional, New York 1993.
Berger P., Luckmann T., *The Social Construction of Reality: A Treatise in the Sociology of Knowledge*, Anchor, New York 1967.
Breuer H.-P., Petruccione F., *The Theory of Open Quantum Systems*, Clarendon Press, Oxford 2006, pp. 166–172.
Cacioppo J.T., Berntson G.G., *Social Neuroscience*, Psychology Press, New York 2005.
Campbell N.A., Reece J.B., *Biology*, Benjamin Cummings, New York – San Francisco 2005.
Carroll S., *From Eternity to Here: the Quest for the Ultimate Arrow of Time*, Dutton, New York 2010.
Chiao J.Y., *Cultural neuroscience*, "Progress in Brain Research" 2009, no. 178, pp. 287–304.
Copeland B.J., *The Essential Turing*, Clarendon Press, Oxford 2004.
Crutchfield J.P., *Between order and chaos*, "Nature Physics" 2011, no. 8, pp. 17–24.
Damasio A.R., *Descarte's Error*, Harper Collins, New York 2000.
Deutsch D., *The Fabric of Reality: The Science of Parallel Universes and Its Implications*, Penguin Books, London 1998.
Dodelson S., *Modern Cosmology*, Academic Press, New York 2003.
Ellis G.F.R., *Cosmology and local physics*, "Int. J. Mod. Phys." 2002, no. A 17, pp. 2667–2672.
Ellis G.F.R., *True complexity and its associated ontology*, [in:] *Science and Ultimate Reality: Quantum Theory, Cosmology and Complexity*, eds. J.D. Barrow, P.W. Davies, C.L. Harper, Cambridge University Press, Cambridge 2004, pp. 607–636.
Ellis G.F.R., *On the nature of causation in complex systems*, "Transactions of the Royal Society of South Africa" 2008, no. 63 [Centenary Issue], pp. 69–84.

Ellis G.F.R., *Top down causation and emergence: some comments on mechanisms*, "Journ. Roy. Soc. Interface Focus" 2012, no. 2, pp. 126–140.

Ellis G.F.R., *On the limits of quantum theory: contextuality and the quantum-classical cut*, "Annals of Physics" 2012a, no. 327, pp. 1890–1932.

Feynman R.P., *The Character of Physical Law*, Penguin Press 1960, pp. 124–125.

Frith C., *Making Up the Mind: How the Brain Creates Our Mental World*, Blackwell, Malden 2007.

Gilbert S., Epel D., *Ecological Developmental Biology*, Sinauer 2009.

Goodstein, D.L., *States of Matter*, Dover, New York 1985.

Hartmann S., *Effective field theories, reductionism and scientific explanation*, "Studies In History and Philosophy of Science" 2001, Part B, no. 32, pp. 267–304.

Isham C.J., *Lectures on Quantum Theory*, Imperial College Press, London 1995.

Kandel E.R., *The Art of Insight*, Random House 2012.

Khoury J., Weltman A., *Chameleon fields: awaiting surprises for tests of gravity in space*, "Phys. Rev. Lett." 2004, no. 93 (17): 171104, [arXiv:astro ph/0309300].

Khoury J., Weltman A., *Chameleon cosmology*, "Phys. Rev." D 2004a, no. 69 (4): 044026. [arXiv:astro-ph/0309411].

Kurose J., Ross K., *Computer Networking: A Top-Down Approach*, Addison Wesley 2010.

Lafore R.,. *Data Structures and Algorithms in Java*, SAMS 2002.

Laughlin R.W., *Fractional quantisation*, "Rev. Mod. Phys." 1999, no. 71, pp. 863–874.

MacCormick J., *9 Algorithms That Changed the Future*, Princeton University Press, Princeton 2012.

Murphy N., Ellis G.F.R., *On the Moral Nature of the Universe*, Fortress Press, Minneapolis 1995.

Noble D., *A theory of biological relativity: no privileged level of causation*, "Interface Focus" 2012, no. 2, pp. 55–64.

Page M.J. et al., *The suppression of star formation by active galactic nuclei*, "Nature" 2012, no. 485, pp. 213–216.

Purves D., *Brains: How They Seem to Work*, FT Press Science 2010.

Penrose R., *Cycles of Time: An Extraordinary New View of the Universe*, Knopf, New York 2011.

Schweber S., *Physics, Community, and the crisis in physical theory*, "Physics Today" 1993, no. November, pp. 34–40.

Tanenbaum A.S., *Structured Computer Organisation*, Prentice Hall, Englewood Cliffs, NJ 1990.

Walker S.I., Cisneros L., Davies P.C.W., *Evolutionary transitions and top--down causation*, "Proceedings of Artificial Life" 2012, no. XIII, pp. 283–290 [arXiv:1207.4808].

Walker S.I., Davies P.C.W., *The Algorithmic Origins of Life*, 2012, arXiv:1207.4803.

Ziman J.M., *Principles of the Theory of Solids*, Cambridge University Press, Cambridge 1979.

Anderson P.W., *More is different*, "Science" 1972, no. 177, pp. 393–396.

Appendix

The key idea is that of *functional equivalence classes*. Whenever you can identify existence of such equivalence classes, that is an indication that top-down causation is taking place. Indeed this is essentially the ontological nature of the higher level effective entity: a computer program is in its nature the same as the set of all possible implementations of the set of logical operations it entails. These are what enter into the higher level effective relations; they can be described in many different ways, and implemented in many different ways; what remains the same in those variants is the core nature of the entity itself.

An *equivalence relation* is a binary relation ~ satisfying three properties:

- For every element a in X, $a \sim a$ (reflexivity),
- For every two elements a and b in X, if $a \sim b$, then $b \sim a$ (symmetry), and
- For every three elements a, b, and c in X, if $a \sim b$ and $b \sim c$, then $a \sim c$ (transitivity).

The *equivalence class* of an element a is denoted $[a]$ and may be defined as the set of elements that are related to a by ~.

The way this works is illustrated in Figures 1 and 2. Higher level states H_I can be realised via various lower level states L_I. This may or may not result in coherent higher level action arising out of the lower level dynamics.

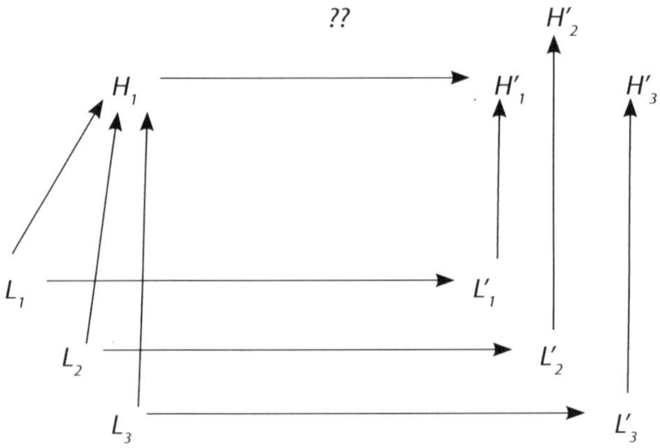

Figure 1

The lower level dynamics does not lead to coherent higher level dynamics when the lower level dynamics acting on different lower level states corresponding to a single higher level state, give new lower level states corresponding to different higher level states.

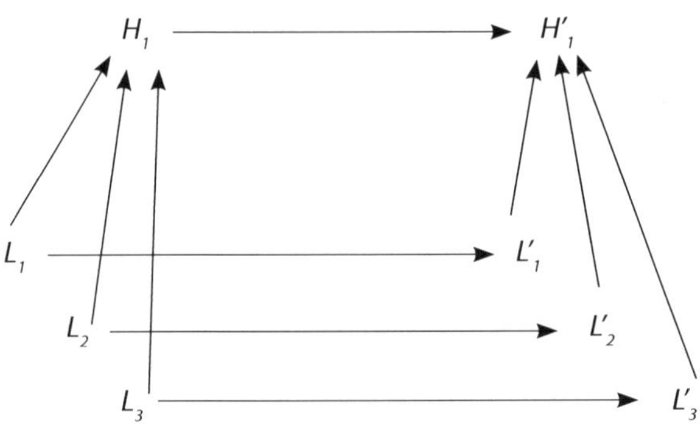

Figure 2

The lower level dynamics leads to coherent higher level dynamics when the lower level dynamics acting on different lower level states corresponding to a single higher level state, give new lower level states corresponding to the same higher level state.

When coherent dynamics emerges, the set of all lower states corresponding to a single higher level state form an equivalence class as far as the higher level actions are concerned. They are indeed the effective variables that matter, rather than the specific lower level state that instantiates the higher level one. This is why lower level equivalence classes are the key to understanding the dynamics.

The ideal gas law: An example is the ideal gas law for gas constrained in a cylinder by a piston:

$$PV = n\,R\,T \tag{8}$$

where the relation between the macroscopic variables pressure *P*, volume *V*, amount of gas *n*, and temperature *T* does not depend on detailed microscopic variables such as velocities, positions and masses of any specific molecules; indeed we don't know those values. The gas is constrained by the cylinder and piston. The variables at the macro level are the only handle we have on lower level states: we can't (except in unusual circumstances) manipulate the micro level variables directly. But we can change them by altering macro variables (e.g. by compressing the gas, so changing *V*), which then changes lower level states (speeding up the molecules); this is the top down effect of the higher level variables on lower level states. The universal constant *R* is the link between the micro and macro states, because it relates the energy of micro states to the values observed at the bulk level.

2. Computer text: the keyboard letters are never exactly identical: yet the abstract letter 'A' represented by them is still the letter 'A' despite all the variations that can occur. It is also the letter 'A' if

- you change font (Times New Roman to Helvetica)
- you change to bold or italic
- you change size of the font
- you change colour of the font
- you change the medium from light on a computer screen to ink on paper

One of the key problems in Artificial intelligence is to assign all these different representations to the same abstract entity that they all represent. This way varied lower level representations of a higher level entity occur is characteristic of top-down causation: what matters is the equivalence class of all these representations, which is the characteristic of the higher level entity, not which particular representation has been chosen. So all those different appearances all represent the same thing. And our minds easily handle this and recognize the higher level abstract thing all these phenomena represent, whether it is the letter 'A' or the plan for a jumbo jet airliner. Those higher level entities (such as the plan for the airliner) really exist as entities in their own right. Proof: Jumbo jet airliners exist. It could not do so unless the abstract plan, with all its multiple representations, were real.

Jean-Philippe Uzan
Université Paris VI Pierre et Marie Curie

Models of the Cosmos and the Emergence of Complexity

1. Introduction

The progress of physical cosmology over the past ten years has led to a 'standard' cosmological model in agreement with all of the available data. Its parameters are measured with increasing precision but it requires the introduction of a dark sector, including dark matter and dark energy, that is attracting the attention of both observers and theoreticians.

Among all the observational conclusions, the existence of a recent acceleration phase of the cosmic expansion is becoming more and more robust. The quest for the understanding of its physical origin is, however, just beginning [2, 3] and it has trigerred questions about the validity of the cosmological model, the robustness of its hypothesis and about the deep nature of reality. Models and speculations are flourishing and we may wonder to what extent the observations of our local universe may reveal the physical nature of dark energy and help reveal the structure of the universe on scales larger than the size of the observable universe. In particular, there exist limitations to this quest intrinsic to cosmology, related to the fact that most observations are located on our past light-cone [4], and to finite volume effects [5] that can make many physically acceptable possibilities undistinguishable in practice.

This text aims at discussing the hypothesis underlying our physical representation of the universe (§ 2) and to illustrate that cosmological data now have the potential to test this hypothesis (§ 3), which goes beyond the more usual measurements of its parameters. This will emphasize the deep connection between theoretical physics and cosmology. I will then discuss (§ 4) the importance of cosmology to understand the emergence of complexity. In this context, the role of effective theories and emergent phenomena is central and I will discuss the notion of causality in this case.

2. Cosmological Models

A cosmological model is a mathematical representation of our universe that is based on the laws of nature that have been derived and validated locally in our Solar system. It thus sits at the cross-road between theoretical physics and astronomy.

While a topic of interest for many centuries, we can safely state that scientific cosmology was born with Einstein's general relativity, a theory of gravitation that made space and time dynamic physical fields that need to be determined by solving dynamical equations known as Einstein field equations. The concept of spacetime was born with this theory. The construction of a cosmological model thus depends on our knowledge of microphysics but also on an *a priori* hypothesis on the geometry of the spacetime describing our universe.

Theoretical physics describes the fundamental components of nature and their interactions. These laws can be probed locally by experiments. They need to be extrapolated to construct a cosmological model. On the one hand, any new idea or discovery can naturally call for an extension of our cosmological model (e.g. introducing massive neutrinos in cosmology is now mandatory). On the other hand, cosmology can help to constrain extrapolations of the established reference theories to regimes that cannot be accessed locally. Let us however note that the knowledge of the laws of microphysics is not sufficient to construct a physical representation of the universe for 2 reasons: (1) the microscopic equations are usually coupled non-linear partial differential equations that cannot be solved in their full generality and require the choice of boundary and initial conditions and (2) solving the Einstein equations requires to describe the nature and distribution of matter, which sets strong limitations since observationally we only have access to the distribution of matter on our own past lightcone on the observable universe, i.e. on a single and limited 3-dimensional slice of the 4-dimensional spacetime. Besides, in practice, we only have access to some of the matter distributed on this past lightcone, because of selection effects. These are reasons for the need of extra-hypothesis, that we call cosmological hypothesis and it raises the question of the representativity of our local Hubble volume compared to the whole universe.

Astronomy confronts us with phenomena that we have to understand and explain consistently. This often requires the introduction of a hypothesis beyond those of the physical theories in order to '*save the phenomena*' [6], as is actually the case with the dark sector of our cosmological model. Needless to say, that even if a cosmological model is in agreement with all observations, whatever their accuracy, it does not prove that it is the 'correct' model of the universe, in the sense that it is the correct cosmological extrapolation and solution of the local physical laws.

When confronted with an inconsistency in the model, one can either invoke the need for new physics or a modification of the laws of physics we have extrapolated in a regime outside of the domain of validity that has been established so far (e.g. large cosmological distance, low curvature regime etc.), or have a more conservative attitude concerning fundamental physics and modify the cosmological hypothesis.

It follows that the construction of any cosmological model relies on 4 main hypotheses (see Ref. [1] for a detailed description),

(H1) a theory of gravity,
(H2) a description of the matter contained in the universe and their non-gravitational interactions,

(H3) symmetry hypothesis,
(H4) an hypothesis on the global structure, i.e. the topology, of the universe.

These hypotheses are not on the same footing since H1 and H2 refer to physical theories. These two are however insufficient for solving the field equations and we must make an assumption on the symmetries (H3) of the solutions describing our universe on large scales while H4 is an assumption of some of the global properties of these cosmological solutions, with the same local geometry.

Our reference cosmological model is known as the ΛCDM model. It assumes that gravity is described by general relativity (H1), that the universe contains the fields of the standard model of particle physics plus some dark matter (DM) and a cosmological constant (Λ), the latter two having no physical explanation at the moment. Their existence is infered from cosmological observations assuming the validity of general relativity (e.g. flat rotation curves, large scale structure, dynamics of galaxy clusters for dark matter, accelerated cosmic expansion for the cosmological constant; see chapters 7 and 12 of Ref. [2]). Note also that this involves an extra-assumption since what will be required by the Einstein equations is the effective stress-energy tensor averaged on cosmological scales. It thus implicitly refers to a, usually not explicit, averaging procedure. It also deeply involves the Copernican principle as a symmetry hypothesis (H3), without which the Einstein equation usually cannot be solved, and usually assumes that the spatial sections are simply connected (H4). H2 and H3 imply that the description of standard matter can be reduced to a mixture of a pressureless and a radiation perfect fluids but, more important that it is a continuous distribution (see Ref. [28] for a detailed description of this hypothesis).

From hypothesis H3, one concludes that the universe must be dynamic and that space is in dilatation (or expansion). Among the robust conclusions of this model is the fact that the universe emerges from a hot, dense and very homogeneous and isotropic phase at thermal equilibrium. The description of the early radiation dominated phase does not depend on any consideration on the dark sector.

The standard history of our universe is summarized in Fig. 1. Interestingly, the evolution of the universe and of its structures spans a period ranging from about 1 second after the Big Bang to today. This is made possible by the fact that (1) the relevant microphysics is well known in the range of energies reached by the thermal bath during that period (<100 MeV typically) and (2) most of the observables can be described by a linear perturbation theory, which technically simplifies the analysis. This description is in agreement with all of the observations performed so far (big bang nucleosynthesis abundances, cosmic microwave background temperature and polarisation anisotropies, distribution of galaxies and galaxy clusters given by large catalogs and weak lensing observations, supernovae data and their implications for the Hubble diagram).

Still, this efficient model lets us with some puzzling open questions.

- *Dark matter*: this pressureless component of matter represents roughly 23% of the total matter content today. It is not included in the model of particle physics as we know it but there exist many candidates in a number of its extensions (such as the lightest supersymmetric particle). It can be optimistically thought that the conjonction of astrophysical observations, direct searches and accelerator experiments shall characterize this particle within the coming ten years. An alternative to dark matter is the formulation of a modified theory for gravity, among which MOND (see chapter 7 of Ref. [2]).
- *Dark energy*: this component is required to explain the late time acceleration of the universe. It represents about 73% of the total matter content today. The required value for the simplest dark energy model, i.e. the cosmological constant, is more than 60 orders of magnitude smaller than what is expected from theoretical grounds (see chapter 12 of Ref. [2]). Hence, dark energy confronts us with a compatibility problem since, in order to

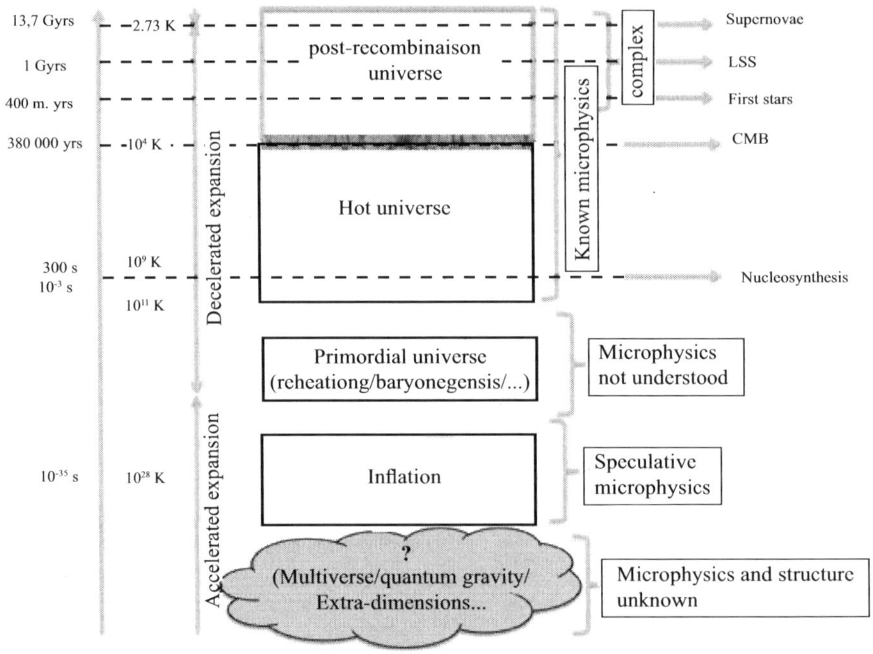

Figure 1

The standard history of our universe. The local universe provides observations on phenomena from the Big Bang nucleosynthesis to today spanning a range between 10^{-3} s to 13,7 Gyrs. One major transition is the equality which separates the universe into two eras: a matter dominated era during which large scale structures can grow and a radiation dominated era during which radiation pressure plays a central role, in particular in allowing for acoustic waves. Equality is followed by recombination, which can be observed through the CMB anisotropies. For temperatures larger that 10^{11} K, the microphysics is less understood and more speculative. Many phenomena such as baryogenesis and reheating still need to be understood. Earlier, the microphysics relies on assumptions that may seem out of reach today, such as the existence of a scalar field driving the dynamics. The whole history of our universe appears as a parenthesis of decelerated expansion, during which complex structures can form, in between two periods of accelerated expansion, which do not allow for thiese complex structures to either appear or even survive.

'save the phenomena' of the observations, we have to include new ingredients (constant, matter fields or interactions) beyond those of our established physical theories [3].

- *Early universe*: the description of the dynamics of the early universe involves a phase of inflation needed to solve the standard problems of the Big Bang model (flatness, horizon, relics; see chapter 8 of Ref. [2]). It has been understood that the quantum behaviour of a scalar field during this period allows us to understand the existence of super-Hubble, almost scale invariant density perturbations. While very successful, the physics during this era is far beyond what can be tested locally. In particular, the nature of the inflaton has to be determined, as well as the robustness of the predictions with respect to the hypothesis on the curvature or isotropy of space.
- *General relativity*: the theory of gravity is extremely well tested in the Solar system but extrapolated in regimes in which we have little control: large distance, low curvature and low acceleration. As explained, controlling the domain of validity of general relativity is crucial in the inference of the existence of the two dark sector components. I refer to [1, 7] for a discussion of the extensions of general relativity and its astrophysical tests.
- *Fundamental constants*: the constants, among which is the cosmological constant, seem to be fine-tuned for a life supporting universe to exist. They are also quantities that cannot be explained by the theories in which they appear. They are at the origin of a double debate: (1) are they really constant?, a question related to the validity of Einstein equivalence principle; see Refs. [8, 9] and (2) can we explain their value?, a question related to the debate on the existence of a multiverse and to the use of anthropic reasoning in cosmology. These two questions actually differ on the scales on which these variations appear, compared to the scale of the observable universe.
- *(in)finiteness of the universe*: while it is impossible to observationally demonstrate that the universe is infinite, we can demonstrate its finiteness in two cases: either by showing that the spatial sections are locally spherical [10] or by revealing a topological structure on a scale smaller than the size of the

observable universe. Such an observational proof would have tremendous implications for cosmology, inflation and the multiverse debate.

- *Copernican principle*: the hypothesis on the geometrical structure of the spatial sections of the universe is central in most conclusions. In particular, the accelerated cosmic expansion can be addressed without invoking dark energy in the framework of inhomogeneous universes. In eternal inflationary models, the universe has a fractal structure on large scales and, in the multiverse picture, the anthropic principle limits the implications of the Copernican principle, which is also an unrealistic assumption when applied to the whole universe, on scales much larger than the Hubble sphere. The main question is thus the scale on which the Copernican principle is expected to hold, and whether its scale is much larger (inflationary case) or of the same order (inhomogeneous cosmology) than the Hubble length.
- *Primordial singularity*: the cosmological model exhibits a spacelike singularity when extrapolated in the past. This Big Bang is not a physical event but a limit in a given theoretical model. It is important to understand how this conclusion may be changed by quantum theories of gravity. It also appears as an explicit limit of our model that motivates many questions on the origin of the universe, that goes beyond the purpose of the model itself.

3. Tests of the Hypothesis

The observational developments now allow us to test the hypothesis of our cosmological model. This will allow us either to find a solution to some of the current puzzles or increase our confidence in the conclusions of our model. Let us summarize where we currently stand.

H4: The debate concerning topology is the continuation of the historically long discussed question of whether our universe is finite or infinite. The hypothesis on global topology does not influence local physics and leaves most of the theoretical and observational conclusions unchanged. Local geometry has, however, a deep impact since it sets the topologies that are acceptable. In our cosmological model, the Copernican principle implies that we are dealing with 3-dimensional spaces of a constant curvature. Besides, almost spatial flatness limits the topologies that can be detectable [12].

A non-trivial topology would violate global isotropy and imprints signatures mainly on the statistical isotropy of the cosmic microwave background anisotropies [11]. The current constraints imply [13] that the size of the universe has to be larger than 0.91 times the diameter of the last scattering surface, that is 24 Gpc. This limit can be ameliorated with the Planck observations. In particular, one can show that it is possible to set constraints on a scale larger than the size of the observable universe, if one restricts himself to a class of topologies [14].

H3: The underlying symmetry hypothesis of our model is the *Copernican principle*, stating that we are not located at a favoured position in space. Combined with the observed isotropy, this leads to a Friedmann-Lemaitre geometry, at least on the scales of the observable universe.

Under this assumption, the spacetime metric is reduced to a single function, the scale factor $a(t)$ that can be Taylor expanded as $a(t) = a_0 + H_0(t - t_0) - \frac{1}{2} q_0 H_0^2 (t - t_0)^2 + \ldots$ where H_0 and q_0 are the Hubble and deceleration parameters. Cosmological observations indicate that $q_0 < 0$: the expansion is accelerating. This conclusion involves no hypothesis about the theory of gravity or the matter content [1], as long as the Copernican principle holds.

The possibility that we may be living close to the centre of a large under-dense region has sparked considerable interest, because such models can successfully reproduce the magnitude-redshift relation of type Ia supernovae without the need to modify general relativity or add dark energy.

A series of tests of the Copernican principle have recently been proposed based either on the CMB spectrum [16] or the time drift of the cosmological redshifts [15]. It was shown that the latter can probe whether our universe is radially homogeneous. In particular it allows us to distinguish models that would otherwise have the same light-cone structure [17]. While challenging, the measurement of the red-shift drift may be achieved with Extremely Large Telescopes and, in particular, it is one of the main drivers of the COsmic Dynamics EX-periment spectrograph [20]. More recently, it has also been proposed [18] that future weak-lensing experiments such as Euclid [19], will allow us to constrain spatial isotropy from the B-modes of the cos-mic shear.

If the Copernican principle holds, the acceleration of the universe can only be accounted for by introducing new fields. Depending on the nature and couplings of these physical degrees of freedom, four universality classes can be defined (see Fig. 1 of Ref. [1] and Ref [3]). Null tests distinguishing between these classes, hence tests of **H1** and **H2**, have also been designed.

H1: Let us define *gravity* as the *long range force that cannot be screened*. Today, it is described by general relativity and is well tested on Solar system scales [21]. We can aim to test both the field equations of general relativity and the equivalence principle, on which Einstein theory relies, on cosmological scales.

The test of local position invariance has been extended to astrophysical scales by studying the constancy of the fundamental constants [8, 9]. Even though such variations may solve some problems (such as the lithium problem [22]) we have no compelling evidence of any variation, which sets strong constraints on the couplings between the dark energy degrees of freedom and ordinary matter.

Extracting constraints on deviations from general relativity is difficult because large scale structures entangle the properties of matter and gravity. On sub-Hubble scales, one can however construct tests reproducing those in the Solar system. For instance, light deflection is a test of general relativity because we can measure independently the deflection angle and the mass of the Sun. This can be generalised to astrophysical scales, as first argued in Ref. [7], since the gravitational potential and matter density perturbation can be measured independently from the use of cosmic shear measurements and galaxy catalogs. This offers a test of the Poisson equation $\Delta\Phi = 4\pi G\delta p$ that has been shown to hold observationally to about 10 Mpc [24]. It is one of the major goals of the Euclid satellite experiment [19]. Many tests of general relativity on astrophysical scales have followed, mainly based on the rigidity between the growth rate of density perturbation and the Hubble expansion, leading to the possibility of designing a post-ΛCDM formalism, sketched in Ref. [3].

H2: A cosmological model has to include all matter fields known locally and take into account any progress (e.g. Higgs or neutrino mass, new particle etc.). For instance, the recent indication [25] of the existence of a particle compatible with the standard Higgs field is an important conclusion that increases the credence we give in the description of the early phase of our model. Besides, the discovery of the first scalar field makes this kind of matter no more speculative, which has deep implications for inflationary models.

One important limitation arises because we need to know the 'average' stress-energy tensor of these matter fields on cosmological scales and our conclusions may be affected by this averaging [31]. The small scale inhomoheneities have been argued to be responsible for a backreaction effect that could explain the late time acceleration [26]. We emphasize that, if correct, this shall also predict that the cosmic expansion becomes anisotropic at a late time [27], a conclusion that may allow us to constrain this mechanism (as discussed just above). More important, the small scale geometry (the fact that matter is not distributed as a continuous fluid on small scales, e.g. of the order of the Solar system size, and that it is composed of collapsed objects and voids) has a drastic impact on the propagation of light. It can significantly impact the Hubble diagram inferred from type Ia supernovae and bias the measurement of the cosmological parameters [28].

As explained earlier, the cosmological constant is the best candidate and is compatible with all of the available data. However, it forces us to face the cosmological constant problem, and most probably to accept embedding our universe in a landscape type of structure, as argued by string theory. All other explanations implicitly assume that this problem is solved and invoke some dynamical component. These models can usually be arbitrarily close to our ΛCDM model, so that they cannot be excluded but are useful to quantify how close from the ΛCDM we must be. Surveys to measure its equation of state w are being planned but the only robust conclusion that we may hope to reach is that $w \neq -1$, that is that the energy density of the dark energy is not constant, hence demonstrating the need to extend our ΛCDM. Other conclusions, such as a best-fit value of w (in some parameterisation), will have only a far fetched relation with fundamental physics (see discussion in Ref. [29] and footnote 4 of Ref. [3]), even though it can sometimes be connected with local constraints [30].

To finish, since any observation involves electromagnetism, we need to test Maxwell theory. This can be achieved through the test of the distance duality relation between angular and luminosity distances [32] that was constructed from the combination of X-ray and SZ observations of clusters. It highly constrains models in which the photon number is not conserved, hence explaining why supernovae appear systematically dimmer. This test shows that no such deviation was likely up to $z \sim 0{,}8$.

The dark sector may be related to the existence of new physical degrees of freedom, and the need for new physics seems compelling from a theoretical point of view. It is however, a triviality to recall that the theory cannot be inferred from observations (see Ref. [3] for the reconstruction of various models reproducing the same data). Indeed, as proven many times, cosmology can exclude extrapolations of our laws of physics and new models. However, since we only have a cosmological model, and not a cosmological theory, validated on the local universe, there are intrinsic limitations to our will to actually probe new physics and we have to be careful in our conclusions. In particular, the line between simple solutions of general relativity with new physics and geometrically more complicated solutions with standard physics has to be drawn.

The tests summarized in this section are *null tests*, in the spirit of 'traditional' physics in which a reference model is confronted with observations in order to determine/extend its domain of validity. Most of them have been or can be performed with observations that were not planned for this purpose. They can be used hierarchically to investigate the dark sector (see Fig. 2 of Ref. [3]). Not denying the use of cosmology to test high-energy physics and the need to consider the extrapolations of these theories (but not simply as a justification to an ad hoc phenomenology), these tests may tone down our claims on what can be proven observationally and help us know how likely it is for cosmological data to unveil some of the physical nature of the dark sector. Any deviation to one of these tests would give a hint to extend our reference cosmological model. No doubt cosmology, with its growing observations, and these tests can teach us more about general relativity and its solutions and, at least, determine how close to a pure ΛCDM the model of our universe must be. With no sign at all, we may have to accept the landscape-type of hypothesis to make sense of our observable universe.

4. The Emergence of Complexity

Section 2 shows that the phenomena that can be observed in our local universe, from BBN time to now, rely mostly on general relativity, electromagnetism and nuclear physics. This is an important issue since the fact that we can understand our universe does not rely on our ability to construct e.g. a theory of quantum gravity (this is the positive aspect, the negative one being that it may be very difficult to find systems in which gravity, quantum physics and observations appear together, inflationary perturbations being probably the only one).

The fact that we can understand the universe and its laws also has another implication on the structure of the physical theories. At each step in our construction of physical theories, we have been dealing with phenomena below a typical energy scale, for technological constraints, and it turned out (experimentally) that we have always been able to design a consistent theory valid in such a restricted regime. This is not expected in general and is deeply rooted in the mathematical structure of the theories that describe nature.

We can call such a property a *scale decoupling principle* and it refers to the fact that there exist energy scales below which effective theories are sufficient to understand a set of physical phenomena that can be observed. *Effective theories* are the most fundamental concepts in the scientific approach to the understanding of nature and they always come with a domain of validity inside which they are efficient to describe all related phenomena. They are a successful explanation at a given level of complexity based on concepts of that particular level. For instance, we do not need to have understood and formulated string theory to formulate nuclear physics and we do not need to know anything about nuclear physics to develop atomic physics or chemistry, needless to say biology. Similarly, electromagnetism describes how protons and electrons attract each other independently of the fact that we have understood, or not, that protons are made of quarks and gluons. This implies that the structure of the theories are such that there is a kind of stability and independence of higher levels with respect to more fundamental ones. It follows that various disciplines have developed independently in almost quasi-autonomous domains, each of them having its own ontology and dynamics that are independent of our ability to formulate a theory explaining these concepts. Sometimes two such effective theories will collide and show an inconsistency that will need, in order to be resolved, the introduction of new concepts, more fundamental, from which the concepts of each of the theories can be derived in a limiting behaviour. For example, Maxwell electromagnetism and Galilean kinematics are incompatible at high velocity, which is at the origin of special relativity with the new concept of spacetime, or the concept of wave function from which the preexisting concepts of particle and wave are limiting behaviours. Note that it implies that concepts that were thought to be incommensurable (such as space and time, or momentum and wave number) need to be unified, which is usually achieved by the introduction of new fundamental constants (speed of light, or Planck constant, in the two examples at hand) that were not considered as fundamental (or even existed) in the previous theories (see e.g. Ref. [34] for a full description of the role of constants in that context).

It follows that the set of theories we are using to describe the world around us can be split into a hierarchy of levels of reality (see Fig. 2, left) as characterised by their corresponding academic subjects. We emphasize that this hierarchy is based on a hierarchy of explanations (space science is 'needed' to do astronomy). It is however different to what happens dynamically since the formation of planets requires the first stars to have existed. Note also that the right branch is related, from a spacetime point of view, to the left since one needs stars/planets for life to develop (there is hierarchy in the formation of structure).

Higher level behaviours are constrained from lower level laws from which they emerge. The relation between the different levels have the following properties (the ways the different levels of complexity interact has been studied in depth by George Ellis [33]):

- Higher level behaviours are constrained by the lower level laws from which they emerge. This is the usual bottom-up causation in which microscopic forces determine what happens at the higher levels. The more fundamental level provides the space of possibilities in which a higher level can develop, by constraing e.g. causality, the type of interactions or structures that can exist. As an example in nuclear physics, a free neutron is unstable and will decay into a proton. While a fundamental process of nuclear physics, it can be understood from the quark composition of the neutron and proton why a neutron decays into proton and not the reverse.

- Scale separation implies that at each level of complexity, one can define fundamental concepts that are not affected by the fact that they may not be fundamental at a lower level (e.g. we can consider proton and neutron as fundamental particles to describe many nuclear properties and forget that they are made of quarks). In that sense, many of the higher level phenomena remain quite independent of the microscopic structures, fields and interaction.
- At least for the lowest level, the fact that physical theories are renormalisable implies that they influence the higher level mostly through some numbers, especially in the case of the fundamental constants of a given effective theory. They cannot be explained within the framework of this particular theory since there exists nothing more fundamental. They can however be replaced by more fundamental constants of an underlying level. For instance, the mass or the gyromagnetic factor of the proton are fundamental constants of nuclear physics. They can however be 'explained', even if the actual computation may be difficult (see Ref. [35]), in terms of constants of the lower level (such as the quark mass, binding energies). This explanation of the constants of an effective theory may reveal new phenomena that cannot be dealt with before (e.g. the fundamental parameters of the effective theory may now be varying) but these phenomena have to be at the margin (or below the error bars) of the experiences that have validated the effective theory, since otherwise it would not have been a good theory!

- Not all the concepts of a higher level can be explained in terms of the lower levels concept. Each level may require its own concepts that do not exist, or even are not related, to lower level concepts. These are emergent properties so that the whole may not be understood in terms of its parts. A typical example is the notion of information that cannot be reduced to chemistry or physical concepts. The concept of temperature, however, can be explained in terms of lower level concepts such as mean kinetic energy.
- The fact that there exists a lower level of complexity and thus microscopic degrees of freedom implies that these degrees of freedom can be heated up so that we expect to observe the apparition of entropy and dissipation. Black hole entropy and Hawking temperature are one argument for the fact that gravity may not be a fundamental interaction.
- The higher levels of complexity can backreact on the lower level. This is the notion of top-down causation. It can take different from such as contextuality, selection effect, control loop etc (see Ref. [33]). Note that the understanding of phenomena of the whole, i.e. of a higher level of complexity, contributes to the understanding of the properties and dynamics of the parts as they function in, and also allow for the existence of, the global structure.

The universe is, at the moment, the largest embedding structure in which all phenomena of lower complexity levels, do develop. It is thus important to describe how the structure of the universe and its history offers the possibility for the emergence of other levels. The cosmological model describes an expanding universe cooling down and developing non-linear structures through the action of gravity, from an initial state at thermal equilibrium. We can thus summarize the evolution of our universe as (hot, homegenous, simple) → (cold, inhomogeneous, complex). In this particular case, most of the history of the universe relies on the property of general relativity. The fact that it is universal and the equivalence principle tells us that the microscopic nature of the matter is irrelevant to determine the large scale structure of spacetime since all that matters is the averaged stress-energy tensor on the scales of interest.

Level 10:	Cosmology	Sociology/Economics/Politics
Level 9:	Astronomy	Psychology
Level 8:	Space science	Physiology
Level 7:	Geology, Earth science	Cell biology
Level 6:	Materials science	Biochemistry
Level 5:	Physical Chemistry	Chemistry
Level 4:	Atomic Physics	Atomic Physics
Level 3:	Nuclear Physics	Nuclear Physics
Level 2:	Particle physics	Particle physics
Level 1:	Fundamental Theory	Fundamental Theory

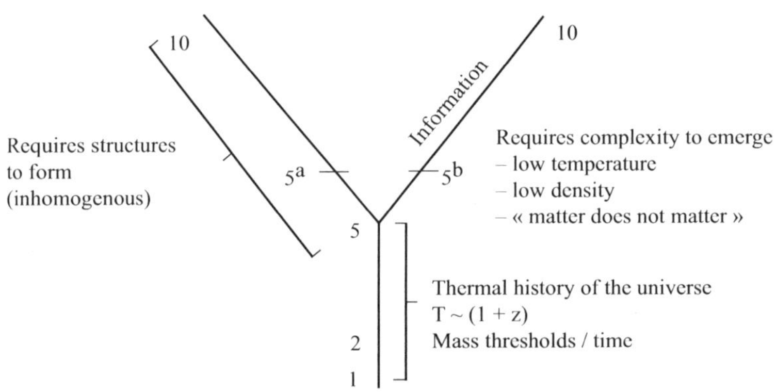

Figure 2

Hierarchy of theories in terms of their level of complexity as proposed by G.F.R. Ellis [33] (previous page) and rephrased (up) to make explicit the bifurcation that appears at the level 5 of complexity at which one needs to distinguish mineral chemistry (5a) and organic chemistry (5b). At this transition a central emergent quantity, *information* appears on the left branch leading to new phenomena such as reproduction-selection, evolution, life and consciousness. The development of the complexity levels of the right branch are however conditional to those of the left branch.

In the early universe, higher levels of complexity cannot emerge, simply because the photon thermal bath is too energetic for complex structures to be stable. For instance, there cannot be any light nuclei before BBN so that at temperature above 100 MeV there exist only free proton's, neutron's, neutrinos, photon's, electron's and positron's at thermal equilibrium under the action of the weak and electromagnetism interactions (see chapter 4 of Ref. [2]). Above 100 MeV, nuclear physics is not needed to describe the universe. In the earlier phases, prior to preheating, the matter content of the universe consisted of a scalar field, the inflaton, so that even particle physics was irrelevant. At preheating, one needs to have a description of particle physics to describe the universe (in particular one needs to know the number of relativistic degrees of freedom, the mass thresholds, their coupling to the inflaton itself etc...). Before recombination, the matter in the universe remains ionised so that there is atomic physics going on (even though atomic transitions are important to describe the proccess of recombination) and there cannot be any molecules at that stage since they would instantaneously be destroyed. To start any chemistry (level 5), one needs to wait for the first stars to synthetize elements heavier than the existing hydrogen, helium, beryllium and lithium. The transition to level 5 thus requires structures such as galaxies and the first stars to start forming (in fact one even needs to wait more since these heavier nuclei have to be released back into the instellar medium). This is summarized in Fig. 3.

It follows that no real complexity can emerge in the universe before it has cooled enough and before structures start forming. We can estimate that before a redshift of ~30 no level of complexity above 5b is actually reached in our universe. The complexity levels of the left branch start developing with the formation of the large scale structure. As shown on Fig. 3 not only do the higher levels of complexity appear only at late time, but also in a very inhomogeneous way since they require the density of matter to be large enough for e.g. molecular interactions to be non-negligible.

There is a feature that comes with the appearance of the different levels of complexity. At each step, only a tiny part of the matter

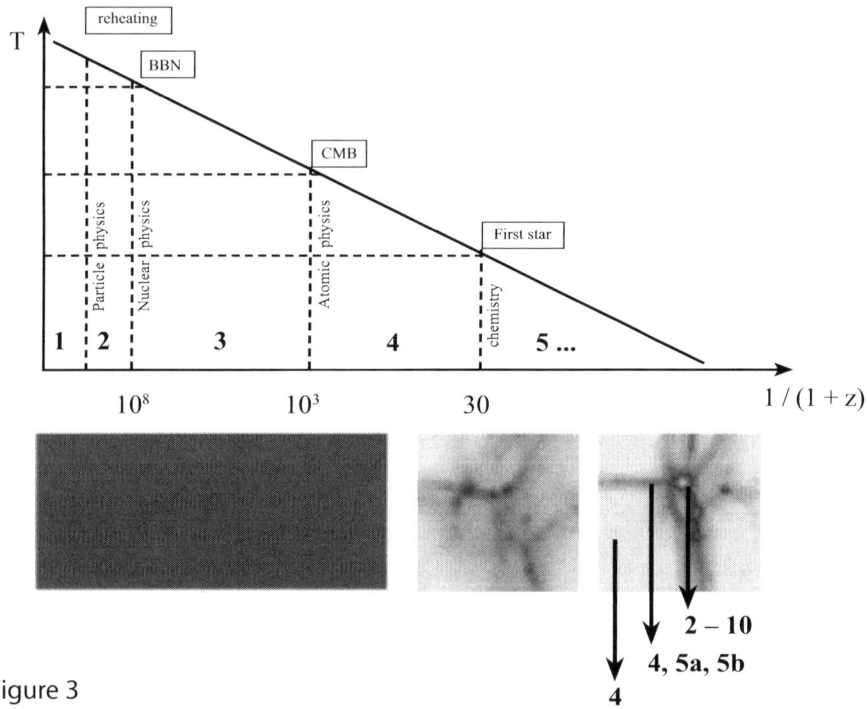

Figure 3

The different levels of complexity presented in Fig. 2 can appear only in regions (and times) where the physical conditions allow for these complex structures, usually more fragile, to be stable. Before decoupling, i.e. 300.000 years after the Big Bang, the temperature was too high for atoms to exist and only very light nuclei were formed so no molecule could have existed. The universe remains mostly homogeneous, hot, with a very simple chemical composition and no level above 4 can emerge. Only after decoupling, do the large scale structures start growing under the effect of gravity (since then radiation pressure has become negligible). The universe becomes inhomogeneous and galaxies, stars and then planets can form locally, allowing (1) for spacetime regions where complexity levels above 5b can be reached and (2) for a more variegated chemical composition with heavier nuclei and then molecules.

content is involved in the next complexity level. This could be referred to as a *decimation principle* [36]. For instance before baryogenesis, matter and anti-matter are in equilibrium. When they annihilate, only a tiny part of the matter content survives and most of it is turned into radiation hence increasing the entropy of the universe.

Then, only a small part of the nuclei are able to participate in chemistry since hydrogen and helium constitute 75% and 24% respectively. Then, only a tiny part of the molecules is organic and only a small part of it contributes to self-reproducing molecules. Whether this is of some relevance and a generic feature of the emergence of higher levels of complexity still needs to be investigated.

This structure of layers of theories of increasing complexity and of nested Russian babushka dolls effective theories also has implications for the notion of causality. George Ellis [33] has described how higher levels can backreact on the properties of the lower level by top-down action. There are however limitations on how it can proceed. Understanding this will teach us what to expect from an explationary point of view from different effective physical theories.

Below level 5, the causality of all theories is inherited from special relativity. Locally they all have a Lorentzian structure so that causality is dictated by the lightcone structure of the embedding spacetime (see Fig. 4-left). The physical conditions at a spacetime point M are dictated by initial conditions within the past lightcone and the equations of evolution. These theories are often asked to have a well-posed Cauchy problem, which ensures that any future field configuration is completely determined by initial conditions and that the notion of 'future configuration' is globally well-defined, which is thought as encompassing the notion of predictability. This property is related to the existence of a fundamental constant which characterizes the maximum speed of propagation of any interaction [37], identified to the speed of light. While this is the case with our current theories, it may not be the case for quantum theories of gravity in which Lorentz invariance may be broken, in some bimetric theories of gravity and even in some more speculative theories in which the whole Lorentzian structure is emergent [38].

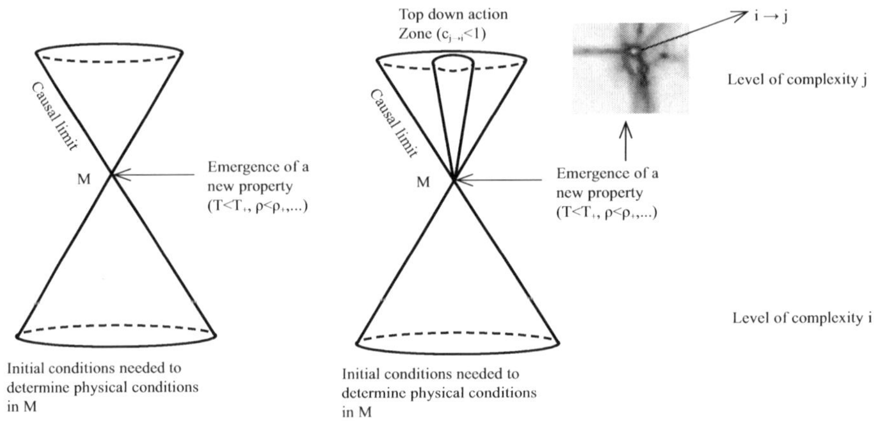

Figure 4

The notion of causality. The physical conditions at a given spacetime event *M* depend on which complexity levels are actually efficient in the universe. Below level 5 (left) the causal structure is determined by special relativity and the lightcone structure, so that the conditions in *M* are unambiguously determined by the evolution of the initial conditions inside the past lightcone. *M* can then influence the conditions within its future lightcone. When a higher complexity level becomes efficient (right), it can backreact on the lower levels. This causation can indeed not extend ouside the future lightcone. Its own effective lightcone is in general much narrower.

Above level 5, top-down action becomes efficient. In general, it implies that the conditions at a given spacetime position become mostly independent of the initial conditions [33]. As argued above, the appearance of these levels of complexity is localized both in time and space, so that the region of the universe in which the top-down causation is efficient is limited. Indeed, it has to be limited by the usual relativistic causality but it is expected that the typical propagation speed of the action of the higher levels on the lower ones is much smaller than the speed of light, so that the future lightcone of influence is expected to be very narrow. For example, in principle humanity can backreact on stellar physics to increase the lifetime of the Sun once it is understood that it can be performed by homogeneizing Solar fuel. It is however clear that the time to achieve the required technological evolution is much larger than 8 minutes. In the same way, the simple fact that the first artificial probes have only recently left the Solar system demonstrates that the domain of influence of (human) life remains restricted to the Solar system. It is thus safe to neglect it for the evolution of our galaxy. Note, however, that it may well modify planetary science since terraformation of Mars is not completely beyond our reach.

5. Conclusions

The standard cosmological model is extremely efficient. It relies of a small set of reasonable hypotheses but the necessity for a dark sector reveals the need for new degrees of freedom which are either physical or geometrical. We have summarized the attempts to test the hypothesis of the model. In particular, it is important to understand the domain of validity of general relativity, the Copernican principle and whether the universe can be finite.

The thermal history of the universe sets the stage for the understanding of the emergence of higher complexity levels. While it works well for the left branch, for which theory and numerical simulations capture the main properties of the development of structures, the role of information and of the right branch requires a better insight. We have presented an attempt of a spacetime analysis of the emergence of complexity and of the top-down action in the real universe. This allows us to (1) to characterize systems for which the higher levels of complexity (and the right branch) can be neglected and (2) characterize systems for which initial conditions are irrelevant. This is important to understand what is the relevant level of description needed to explain a given phenomenon.

To finish, we can wonder why the physical theories we know enjoy the property of scale separation, which is deeply related to the understandability of the world. This is intriguing, in the same sense as the apparent fine tuning of the fundamental constants. We can conjecture that an anthropic explanation can be given. We may also regard the hypothetical multiverse structure as similar to the distribution of the regions of complexity. We have shown that there are islands where the physical conditions allow for the appearance of complex structures to develop in space and time. In the same sense, the multiverse hypothesis proposes that there exist islands in the space of physical theories where fundamental physics allows for complex physics and thus is the prerequisite for the development of complexity. The central question is indeed how to adress the plausibility of such a hypothesis.

6. Acknowledgments

I would like to thank the Copernicus Center and Michael Heller for their kind invitation to Kraków and for organizing such a lively meeting and George Ellis for our long lasting and stimulating collaboration together with his friendship.

References

[1] Uzan J.-P., *Dark energy, gravitation and the Copernican principle*, [in:] *Dark Energy, Observational and Theoretical Approaches*, ed. P. Ruiz-Lapuente, Cambridge Univ. Press, Cambridge 2010, pp. 3–47, [arXiv:0912.5452]

[2] Peter P., Uzan J.-P., *Primordial Cosmology*, Oxford Univ. Press, Oxford 2010.

[3] Uzan J.-P., *The acceleration of the universe and the physics behind it*, "Gen. Rel. Grav." 2007, no. 39, p. 307.

[4] Ellis G.F.R., *Cosmology and verification*, "Q. Jl. astr. Soc." 1975, no. 16, p. 245.

[5] Bernardeau F., Uzan J.-P., *Finite volume effects for non-Gaussian multi-field inflationary models*, "Phys. Rev." 2004, no. D 70: 043533.

[6] Duhem P., *Sozein ta phainomena: Essai sur la notion de théorie physique de Platon à Galilée*, Vrin, Paris 1908; translated as *Sozein ta Phainomena: An Essay on the Idea of Physical Theory from Plato ti Galileo*, Univ. of Chicago Press.

[7] Uzan J.-P., *Tests of general relativity on astrophysical scales*, "Gen. Rel. Grav." 2010, no. 42, p. 2219; Uzan J-P., Bernardeau F., *Lensing at cosmological scales: A test of higher dimensional gravity*, "Phys. Rev." 2001, no. D 64: 083004.

[8] Uzan J.-P., *The fundamental constants and their variation: Observational status and theoretical motivations*, "Rev. Mod. Phys." 2003, no. 75, p. 403, [arXiv:hep-ph/0205340]; Uzan J.-P., *Variation of the constants and tests of gravity on astrophysical scales*, "AIP Conf. Proc." 2005, no. 736, p. 3, [astro-ph/0409424]; Uzan J.-P., *Fundamental constants and tests of general relativity – theoretical and cosmological considerations*, "Space Sci. Rev." 2010, no. 148, p. 249, [arXiv:0907.3081].

[9] Uzan J.-P., *Varying constants, gravitation and cosmology*, "Living Rev. Rel." 2011, no. 14, p. 2, [arXiv:1009.5514 [astro-ph.CO]].

[10] Uzan J.-P., Kirchner U., Ellis G.F.R., *WMAP data and the curvature of space*, "Month. Not. R. Astron. Soc." 2003, no. 344, p. L 65.

[11] Riazuelo A. et al., *Cosmic microwave background anisotropies in multi-connected flat spaces*, "Phys. Rev." 2004, no. D 69: 103518; Riazuelo A. et al., *Simulating cosmic microwave background maps in multi-connected universe*, "Phys. Rev." 2004, no. D 69: 103514; Uzan

J.-P. et al., *Cosmic microwave background constraints on lens spaces*, "Phys. Rev." 2004, no. D 69: 043003; Luminet J.-P. et al., *Dodecahedral space topology as an explanation for weak wide-angle temperature correlations in the cosmic microwave background*, "Nature" 2003 (London), no. 425, p. 593.

[12] Weeks J., Lehoucq R., Uzan J.-P., *Detecting topology in a nearly flat spherical universe*, "Class. Quant. Grav." 2003, no. 20, p. 1529.

[13] Shapiro J. et al., *Extending the WMAP bound on the size of the universe*, "Phys. Rev." 2007, no. D 75: 084034.

[14] Fabre O., Prunet S., Uzan J.-P. in preparation.

[15] Uzan J.-P., Clarkson C., Ellis G.F.R., *Time drift of cosmological redshifts as a test of the Copernican principle*, "Phys. Rev. Lett." 2008, no. 100: 191303.

[16] Caldwel R.R., Stebbins A., *A test of the Copernican principle*, "Phys. Rev. Lett." 2008, no. 100: 191302; Zhang P., Stebbins A., *Confirmation of the Copernican principle at Gpc radial scale and above from the kinetic Sunyaev Zel'dovich effect power spectrum*, "Phys. Rev. Lett." 2011, no. 107: 041301.

[17] Dunsby P. et al., *How close can an inhomogeneous universe mimic the concordance model*, "JCAP" 2010, 1006: 017.

[18] Pitrou C., Uzan J.-P., Pereira T.S., *Weak-lensing B-modes on all scales as a probe of local isotropy*, "Phys. Rev." 2012, D (2012) in press, [arXiv:1203.6029] .

[19] Laureijs R. et al., *Euclid definition study report*, "ESA-SRE" 2011, no. 12, [arXiv:1110.3193 [astro-ph.CO]] .

[20] Pasquini L. et al., *CODEX: Measuring the expansion of the universe (and beyond)*, "The Messenger" 2005, no. 122, p. 10.

[21] Will C.M., *Theory and Experiment in Gravitational Physics*, Cambridge University Press, Cambridge 1981.

[22] Coc A. et al., *Coupled variations of fundamental couplings and primordial nucleosynthesis*, "Phys. Rev." 2007, no. D 76: 023511.

[23] Uzan J.-P., Bernardeau F., *Lensing at cosmological scales: A test of higher dimensional gravity*, "Phys. Rev." 2001, no. D 64: 083004.

[24] Doré O. et al., *Testing Gravity with the CFHTLS-Wide Cosmic Shear Survey and SDSS LRGs*, [arXiv:0712.1599] .

[25] ATLAS Collaboration, *Observation of a new particle in the search for the Standard Model Higgs boson with the ATLAS detector at the LHC*, [arXiv:1207.7214]; CMS Collaboration, *Observation of a new*

boson at a mass of 125 GeV with the CMS experiment at the LHC, [arXiv:1207.7235].

[26] Buchert T., *On average properties of inhomogeneous fluids in general relativity. 1. Dust cosmologies*, "Gen. Rel. Grav." 2000, no. 32, p. 105; Buchert T., *On average properties of inhomogeneous fluids in general relativity: Perfect fluid cosmologies*, "Gen. Rel. Grav." 2001, no. 33, p. 1381.

[27] Marozzi G., Uzan J.-P., *Late time anisotropy as an imprint of cosmological backreaction*, "Phys. Rev." 2012, D, in press, [arXiv:1206.4887].

[28] Clarkson C. et al., *(Mis)-interpreting supernovae data in a lumpy universe*, "Month. Not. R. Astron. Soc." 2012, in press, [arXiv:1109.2484].

[29] Schimd C. et al., *Tracking quintessence with cosmic shear – Constraints from VIRMOS-Descart and CFHTLS and future prospects*, "Astron. Astrophys." 2007, no. 463, p. 405.

[30] Martin J., Schimd C., Uzan J.-P., *Testing* $w < -1$ *in the Solar System*, "Phys. Rev. Lett." 2006, no. 96: 061303.

[31] Ellis G.F.R., *Patchy solutions*, "Nature" 2008, no. 452, p. 158.

[32] Uzan J.-P., Aghanim N., Mellier Y., *The distance duality relation from X-ray and SZ observations of cluster*, "Phys. Rev." 2004, no. D 70: 083533.

[33] Ellis G.F.R., *Physics, complexity and causality*, "Nature" 2005, no. 435, p. 743; Ellis G.F.R., *On the nature of causation in complex systems*; Ellis G.F.R., *Top-down causation and emergence: Some comments on mechanisms*; Ellis G.F.R., *Physics and the real world.*

[34] Uzan J.-P., Lehoucq R., *Les Constantes Fondamentales*, Belin, Paris 2005.

[35] Luo F., Olive K., Uzan J.-P., *Gyromagnetic factors and atomic clock constraints on the variation of fundamental constants*, "Phys. Rev." 2011, D 84: 096004.

[36] I thank Elisabeth Vangioni for pointing to me the importance of this phenomena.

[37] Ellis G.F.R. and Uzan, J.-P., *c is for the speed of light, isn't it?*, "Am. J. Phys." 2005, no. 73:, p. 240.

[38] Mukhoyama, S. and Uzan, J.-P. (2013), *From configuration to dynamics – emergence of Lorentz signature in classical field theory*, "Phys. Rev." 2013, no. D 87:065020.

Derek Raine
University of Leicester

Causality and Complexity

In linear systems there is a direct link between cause and effect. This is not the case in non-linear systems where causes are inter-related. This is, of course, well known in social systems where the attribution of blame is often a matter of opinion or convenience. Often the approach taken is to appeal to a notion of *perturbative causality* where small changes in causes are traced to their effects, other things being equal, or at least, as equal as possible given the constraints. In this situation, small causes may appear to generate large effects at tipping points, neglecting the interaction of the small perturbation with the other system variables. In complex systems it is the interactions of the agents, rather than the attributes of the individual agents, that lead to emergent behaviour, thereby undermining a naïve reductionist view.

In cosmology the emergent properties of the Universe are attributed to special values of the fundamental constants on the grounds that small perturbations of these values cause large deviations from the observed features, particularly the existence of sentient observers. Thus we are not surprised to find ourselves inhabiting an instance of the multiverse that can support life. This is an example of perturbative causality and gives rise to the idea that the parameters of the system have been fine-tuned.

An alternative view seeks to understand emergent behaviour as a probable consequence of complexity. One example is Kauffman's *NK* model applied to the origin of life: in this model life (in the sense of a metabolic network) evolves wherever complex chemistry is possible as the natural outcome of a complex system without any fine tuning. Taking this further, we can imagine that the multiverse admits complex behaviours in most of its ground states. In this view there is nothing special about the values we find for the fundamental constants: life, as a complex system, is bound to emerge in any universe with a sufficient number of interacting fields, that is, in high enough dimensions. I end by questioning whether the notion of top down causation adds anything to this discussion.

1. Introduction

I want to begin with an imaginary awards ceremony, for the cause of the First World War: we will not be deciding the awards, but we shall be looking at some of the nominations that have been proposed in the literature on the conflict. There are too many to list exhaustively, but here are some of them. (I have extracted this list from the Wikipedia article[1]).

In the individual category the sole nomination is for Gavrilo Princip for the murder of the Archduke Ferdinand and his wife, although no doubt Gavrilo will want to thank his support team of Bosnian Serb military.

In the category of countries, Germany has four nominations for various attributes including the inflexibility of its railway timetables; France has one nomination for playing domestic politics; and Britain has a nomination for diplomatic bumbling in its dealings with Europe. (We continue to be good at that).

There is a joint award with nominations for Britain and Germany for indulging in an unstable arms race, to financiers for competing for markets and finally to all participants more or less equally for whatever they did or did not do.

---

[1] http://en.wikipedia.org/wiki/Causes_of_World_War_I; accessed 26/09/2012.

And apparently abstract nouns can cause wars too; so we have in this category: imperialism, nationalism, militarism and social Darwinism.

So what was the cause of the First World War? This is clearly an unanswerable question. The difficulty in assigning a cause arises because the situation is non-linear: that is, there is feedback. In a linear system if we add causes, we must add their effects: if a causes A and b causes B, then a and b acting together cause the outcomes A and B together. Once we introduce interactions, hence non-linearity, either between different parts of a system or self-interaction, this simple link between cause and effect is lost. (The book *Freakonomics* by Levitt & Dubner (2005) has lots of social examples).

2. Perturbative causality

A common resolution, since we seem to be genetically programmed to seek causes for whatever happens, is to try to identify perturbative causes at tipping points. There may be many inputs that bring a situation to the brink, but we like to think of the cause of the subsequent fall-out as the perturbation that took us over the edge. For the First War this is popularly identified as the murder in Sarajevo, but it is not at all clear that perturbative causes can always be assigned correctly by tradition or popular opinion.

Another often quoted example of a perturbative cause is the rejuvenation of Time Square in New York[2]. The history of its development is fascinating, but the main point is the decision of the Disney Corporation to set up a cinema in what had become a run-down, crime infested neighbourhood to avoid. This sparked a raft of other businesses to set up shop, thereby transforming the area. Disney was a tipping point. Or was the tipping point the second business that opened?

---

[2] http://www.nytimes.com/2010/12/04/nyregion/04square.html?pagewanted=all&_moc.semityn.www; accessed 26/09/2012.

Or the third? Of course, we cannot rerun events with different parameters to find out; so today we cannot reliably identify the single cause.

In brief, a perturbative cause may amount to little more than a matter of opinion.

3. Synergistic causality

Another type of causality might be labelled synergistic causality. Turing patterns are a good example. The reaction-diffusion equations for concentrations C_1 and C_2

$$\frac{\partial C_1}{\partial t} = R_1(C_1, C_2) + D_1 \nabla^2 C_1$$

$$\frac{\partial C_2}{\partial t} = R_2(C_1, C_2) + D_2 \nabla^2 C_2$$

and judicious choice of reaction functions R_1 and R_2, can lead to the spotted body and striped tail of the cheetah, the striped body and striped tail of the tiger and the spotted body and spotted tail of the leopard, but not the non-existent striped body and spotted tail (Gierer and Meinhardt 1972). The more rapidly diffusing morphogen suppresses the production of the other, but neither separately is the cause of the patterns. The key point is that it is the interaction of the morphogens that causes the patterns to arise. Furthermore, for these spatial patterns to arise, diffusion has to operate on two different time scales. One might say that the two different time scales of diffusion, given by the ratio of diffusion constants, become visible in the spatial patterns.

4. Criticality

Equally interesting are patterns which arise from random fluctuations at a critical point because these are often cases where we find it difficult to refrain from attributing a non-random cause. The well-known

Ising model provides an archetypal example (see e.g. Bar-Yam 2003, Gros 2010). The model consists of interacting binary elements (quantum mechanical spins in the original physics context) which can take values ±1, arranged on a lattice. An element prefers to align with the same values as its neighbours, in the sense that alignment minimises the interaction energy. But the model assumes also that the alignments are subject to random fluctuations, or noise, (thermally induced in the physical picture). At the boundary between misaligned spins, energy is reduced if a spin is flipped. So if there is not much random noise to undo the arrangement, the spins will all tend to align one way or the other (all −1 or all +1). With enough noise the interaction energy will become irrelevant and the system will be a random mixture of 1s and −1s with small fluctuations which are continually smoothed out. Somewhere in between there is a point where a small fluctuation is not smoothed out but can grow. This is a critical point. The system becomes islands of 1s in a sea of −1s or vice versa – and can flip between the two. The figure shows a plot of a simulation of the average spin against time step.

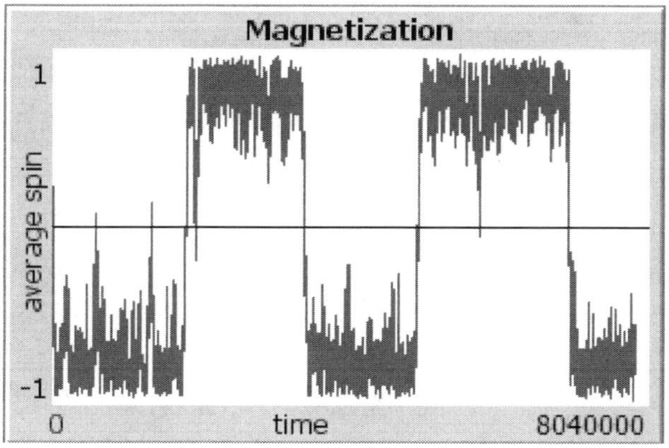

Figure 1

A simulation of the 2d Ising model showing the random switching of magnetisation at the critical temperature. The model comes from the Netlogo library (Wilensky 2003).

It looks as if something is causing the system to flip from one state to the other at particular times. If this were a stock price no doubt a jump up would coincide with the appointment of a new CEO and the jump down would shortly precede his departure. But the CEO would not be master of either event, which are solely due to the random fluctuations in the system at the critical point. The apparent coordinated behaviour of the spins simply emerges from the nature of the interactions. That is, it is the interaction of the agents that is the cause of the macroscopic behaviour.

The idea of emergence in complex systems is both central and difficult to pin down. Examples are easy however. Perhaps the simplest is the behaviour of water. Wetness, viscosity, and even the phase are not predicable from the observation of a single water molecule which has none of these properties. The bulk attributes can nonetheless be traced to the *interaction* of the water molecules, which is therefore the cause of the behaviour.

A crucial point here is that we do not have to know all the details of the microscopic interactions between agents to be able to investigate the macroscopic behaviour of the system. For example, it is sufficient to reduce the representation of the complex quantum configurations of electrons in a ferromagnetic atom to the alignments of binary elements in order to investigate magnetic properties. In short, we have to model the interactions appropriately, not the detailed nature of the agents.

This has implications for the converse attempts to deduce interactions from global behaviour. The first well-known application of agent-based thinking to sociology by Schelling (1971) is a good example. Modelling the apparent segregation in Chicago in the 1960's Schelling found that a preference for a mere 1 in 3 of an agent's neighbours to be of the same colour resulted in the model in segregation of the population into black and white districts. In other words the observed segregation in the model came about even where the population were happy to have a majority of their neighbours of a different colour. The conclusion is that the simple observation of segregation in the city cannot be used as evidence of colour prejudice in the individual agents. Or to put it another way, the cause of the observed segregation is to be found in the way these agents interact, not in this case in their individual prejudices.

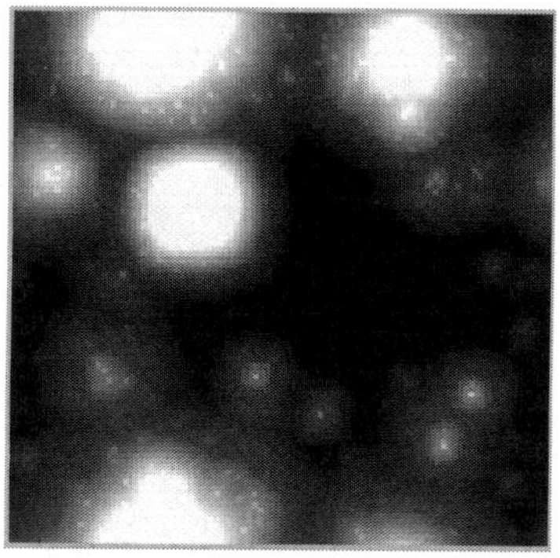

Figure 2

Netlogo model of Angels and Mortals by Iain Weaver from the Netlogo community models (http://ccl.northwestern.edu/netlogo/models/community/Angels and Mortals). The white patches represent surviving mortals.

Finally for this part of the discussion let us look at another example where an agent based view is essential (Shnerb et al 2000). The example we choose is called angels and mortals. The world in this model is initially populated by a uniform cloud of mortals with density n_M. These are treated as continuous, and this cloud diffuses at a rate given by D_M. Next we add a distribution of angels with density n_A. These are treated as discrete, and diffuse at a rate given by D_A. At each time-step a fraction μ of mortals on each finite patch of the world die. However, if there is an angel occupying the patch, the mortals multiply by a factor λ.

We might naively predict that the population will evolve according to the differential equation

$$\frac{dn_B(t)}{dt} = \lambda n_A n_B(t) - \mu n_B(t)$$

with solution

$$n_B(t) = n_B(0)\exp(t(\lambda n_A - \mu))$$

which predicts an exponential decay to extinction for $\lambda n_A < \mu$. In fact, in an agent-based model where we track local interactions, we find that patches of mortals survive for a range of parameters (Fig. 2). In other words, the 'representative agent' makes no sense in this context.

5. Separation of Time Scales

In some of these models the critical behaviour requires tuning of control parameters, for example in the Ising model; in others it is self-organised. As in the well-known sand pile model of self-organised criticality (Bak 1996), the emergence of complexity relies on the separation of time scales (or spatial scales) (Adami 1998, Nicolis and Nicolis 2012). For the sand pile we require that the rate of influx of sand grains tend to 0 in order to manifest complex (power law) behaviour. In the Schelling example we must wait for the effect of each initial decision (move or not move) to ripple through the system before setting up the next move; likewise with the Ising model. In these and the angels and mortals model the time structure becomes visible as spatial structure. In the Turing model we required a separation of diffusion time scales to lead to an emergent spatial patterning.

6. The NK Model

A general framework for adaptive interacting agents that has received a lot of attention is the *NK* model (Kauffman 1993). Here agents are endowed with a 'gene' having a fixed length of N loci at each of which there can be A alleles so there are A^N possible genotypes. The fitness of a genotype depends on the contribution to its fitness from the alleles at each of the loci. These are assigned randomly. If $K = 0$ these individual fitness contributions are just summed. If $K \neq 0$, then the fitness of an allele at each locus depends on fitnesses of the alleles at K neighbouring loci. We can therefore think of the model in terms of a fitness landscape with fitness plotted as a function of genotype in many dimensions. If $K = 0$, the landscape is smooth and single peaked (because flipping the alleles one locus at a time gives a unique maximum fitness). If $K = N - 1$, the landscape is random. (Each choice of alleles is randomly different from any other). For intermediate values of K the landscape has a complex structure.

Evolution can be presented in terms of such a fitness landscape. This framework allows us to visualise adaptive agent behaviour. The fittest agents at each time step reproduce with random mutations, thereby climbing to a fitness peak.

The landscape is not necessarily static – it can itself be a function of the agent variables, in which case it represents co-evolution of the agents. However, once again, adaptation requires a separation of time scales: the evolution of the landscape must be slower than the evolution of the agents.

Taking this one stage further back Kauffman argues that any large set of polymerizing elements will give rise to an autocatalytic network of chemical reactions. In other words, fine tuning is not required for the emergence of metabolic networks. Evolution then proceeds to act on these networks by favouring catalytic efficiency.

7. Examples of Emergence

A couple of our own unpublished models will illustrate emergence in complex adaptive systems. The first example is a pre-biotic ecology in which protocells of different molecular components compete for survival. The molecular structure of the cells determines both their probability of fusion with other cells and whether, if they survive to division, their molecular constituents can catalyse the production of similar or randomly different offspring. For a large fraction of the randomly assigned properties of the molecular constituents, we find that a small number of types of mutually catalytic cells survive to take over the population. Given a vast space of possibilities, the system self-organises to what is effectively a local fitness peak over a timescale long compared with the timescale of agent interactions. (See Raine and Symons (2009) for a basic version of the model).

A second example is an agent-based model of economic growth where companies are represented as agents exploring a technological landscape (Fagiolo and Dosi 2003). Here adding the *NK* model allows us to implement the creative destruction associate with growth by allowing the productivity of old technologies to be dependent, positively or negatively, on newly discovered technologies as agents move randomly through the space. The growth in GDP in the model, as agents exploit the technologies they discover, is enhanced by the imposition of creative destruction. In effect, agents discover the complex web of inter-relations as they adaptively ascend the local fitness peaks. To obtain the emergence of GDP growth, it appears that the timescales for the diffusion of innovation has to be long compared to the periods of exploitation of known technologies.

In both cases, emergent properties arise for a large range of parameters – possibly for almost all parameters – provided that we have a separation of time scales. (Otherwise, the result of random interactions is random behaviour).

8. Fundamental Constants and Life

There are various characterisations of the parameters of a Universe that appear to be tuned to the existence of life. One now quite well-known set is discussed by Rees (2001) which will serve as an illustration. Rees considers the five constants (or parameters) N, ε, Ω, λ, Q and d with the following meanings.

N is the ratio of the strength of electrostatic forces to the strength of gravity as measured for example by the ratio of the electric and gravitational forces between two protons. N is of order 10^{38}. If it were much smaller the lifetime of the Universe would be too short for life to emerge. ε controls the binding of nuclei; its value of 0.007 ensures that the Sun shines and that stars produce the elements of life. $\Omega \sim 1$ measures the amount of matter in the Universe: too high and the Universe collapses under its own gravity; too low and galaxies cannot form. The parameter λ is the cosmological constant; its value is multiple orders of magnitude smaller than one might guess, which is just as well for the sake of the formation of galaxies. $Q \sim 10^{-5}$ is the binding energy of structures in the Universe relative to their rest mass energy. Its value is such as to ensure that low density contrast structures can form. Finally d is the dimensionality of space which turns out to be 3, just as required for stable solar systems to be available for life to form in.

The apparent tuning of all of these parameters gives rise to various forms of the anthropic principle, the weakest statement of which is that we live in a universe fit for observers. How are we to understand this statement?

Consider a solar system that is shrouded from the rest of the Universe so that its cosmologists know of only one star. They might deduce that the formation of the Universe (their one star and its planets) required some very fine tuning to produce such a fine star, so suited to life. More detailed measurements might in time lead them to the detection of a wider universe and the hypothesis that stars are formed in clusters, which are almost certain to contain many planets suitable for life. The existence of life in this wider context is perhaps less of a mystery.

The anthropic principle can therefore be understood in the context of a multiverse. The discovery of an overwhelmingly large number of vacuum states in M-theories (Douglas 2003; Ashok and Douglas 2004) strengthens the multiverse approach (Uzan, 2011). There are nevertheless even in this context two different points of view[3]:

> The first would be to consider that since complexity – and in particular, life – is an adaptive process, it would have emerged in nearly any kind of universe.... The second idea is to accept the existence of many

[3] http://cerncourier.com/cws/article/cern/31860; accessed 26/9/2012.

universes with different laws where we naturally find ourselves in one of those compatible with complexity.

So the question we can ask is: is fine tuning an illusion of our limited characterisation of life and, or, of the perturbative causality we appear to be imposing on the system by varying one parameter at a time? Or, to put it another way, is complexity the norm in most universes. As just one example, the dark energy density λ and the fluctuation amplitude Q are not constrained jointly to observed values by anthropic arguments (Wilczek 2007).

Note that attempts to explore the interior of parameter space are often still couched in terms of the existence of galaxies, or complex nuclei and so on (Bousso et al. 2009, Adams 2008). My point of view is different. Hard as it may be to imagine 'life' (complex adaptive systems) without galaxies or stable planetary systems, for the purpose of this discussion I am interested in the conditions under which complexity might arise, irrespective of whether we understand the form it might take (e.g. Barrow 1993).

As a necessary condition, we have seen that this requires a separation of time scales. However, the crucial point seems to be that a separation of time scales is almost automatic in a classical Universe. From a specific model (Caldeira and Leggett 1985) and more generally (Zurek 1986) the ratio of quantum decoherence time to the relaxation time is given by

$$\frac{\text{decoherence time}}{\text{relaxation time}} = \left(\frac{\lambda}{\Delta x}\right)^2$$

The decoherence time is the time over which quantum correlations are destroyed by interaction with the environment, while the relaxation time is the time over which friction operates in the classical system. The quantity λ is the de-Broglie wavelength ($h/(2\pi\sqrt{2mkT})$) at temperature T and Δx is the separation in the classical system. This ratio is almost always very large.

Universes become classical if there are sufficient interactions between agents, possibly also only if they contain sufficient mass (Raine 1993). On this view, complexity is a (measure 1?) consequence of classicality. Or, to put it another way, observers are (almost?) certain to arise in classical universes. (And we are clearly not concerned with Universes that remain quantum superpositions since nothing happens in those: evolution remains unitary).

9. Top-Down Causation?

So much for mere speculation! The anthropic viewpoint has given rise to a more overarching philosophical position, which has become known as top-down causation. (See for example, Ellis et al. 2012). According to this view the reductionism of classical causality is at odds with emergent phenomena for which a top-down view of causality is required. I find the arguments unconvincing.

First of all, there is no denying that there is more to causality than the Aristotelian 'efficient cause' (the sculptor's chisel on the stone). Top down causality sometimes gives the impression it wants to be a final cause (the purpose of the statue) but seems more often expressed as a formal cause (the vision of the finished sculpture in the mind of the artist).

Staying with artistic endeavour, let us take as an illustrative example of reductionism in the light of the Mona Lisa. It is claimed that on a reductionist view (efficient causes only allowed) that the painting of the picture must be contained in the disposition of the atoms at the beginning of the Universe. This is, of course, absurd. The missing ingredient however is not a formal (or top-down) cause, but chance. The Mona Lisa is an evolutionary adaptation just as is the peacock's tail. Adaptation requires error; that is, chance. Then its interacting agents can climb the fitness landscape. One might argue that the landscape has to be put in place and so its form is the formal cause of evolutionary adaptation. (My impression is that this is Kauffman's view, for example). But the landscape is itself constructed by the interaction of the agents: an alien race with a different physiognomy may well find the Mona Lisa a meaningless jumble, far from a fitness peak.

Of course, I cannot trace in any detail the levels of agency and interaction that produce the Mona Lisa from the big bang. But it is clearly far from proven that any intervention is required to explain emergence in an adaptive (error admitting) physical system.

Now advocates of top-down causality are well aware of all of this. The difference is that they are not satisfied to call it adaptive evolution but call it (lumped together with a number of other things) top-down causality. Calling things by different names is harmless except when it isn't: in particular when it leads to category errors such as the attribution of existence to the thing from the existence of the idea of the thing.

Turning on a switch in an electrical circuit seems to be an example of perturbative causality. But it does not cause the light bulb to glow: it allows the battery to cause the bulb to radiate. It is the result of the interacting agency of the operator of the switch and the battery. Perhaps one might argue that the operation of the light switch is meaningful whereas the battery electrons follow the blind laws of physics. Meaningfulness, like all information, requires a context, and the context creates the meaning; but it is not the cause of the message.

One might argue that context and hence meaning (or information) is embedded in software (in the brain in this case). But software is physical (albeit an equivalence class of physical representations): thoughts are the peacock's feathers, a product of adaptive evolution: electrons flowing through circuits given significance by the interaction with other agents as a result of their mutual evolutionary adaptation.

In this forum I can do no more than raise these doubts, for what they are worth, as to the value of labelling such causes in complex systems as top-down.

References

Adami C., *Introduction to Artificial Life*, Springer 1998.

Adams F.C., *Stars in other universes: stellar structure with different fundamental Constants*, JCAP 2008, doi:10.1088/1475-7516/2008/08/010.

Ashok S., Douglas M., *Counting flux vacua*, JHEP 2004, 0401, 060.

Bak P., *How Nature Works: The Science of Self-organized Criticality*, Springer 1996.

Bar-Yam Y., *Dynamics of Complex Systems* (*Studies in Nonlinearity*), Westview Press 2003.

Barrow J., *Anthropic principles in cosmology*, [in:] *Vistas in Astronomy*, eds. P. Beer, A.E. Roy, R.E. White, "Quantum Physics and the Universe Special Volume" 1993, no. 37 (1–4), pp. 409–428 Pergamon.

Bousso R., Hall L.J., Nomura Y., *Multiverse understanding of cosmological coincidences*, "Phys. Rev." 2009, no. D 80: 063510.

Caldeira A., Leggett A., *Influence of damping on quantum interference: An exactly soluble model*, "Phys. Rev.", no. A 31, pp. 1059–1066.

Douglas M., *The statistics of string / M theory vacua*, JHEP 2003, 0305, 46.

Ellis G.F.R. et al. in Interface Focus 2012 **2** (1) Royal Society Publishing http://rsfs.royalsocietypublishing.org/content/2/1

Fagiolo G., Dosi G., *Exploitation, exploration and innovation in a model of endogenous growth with locally interacting agents*, "Structural Change and Economic Dynamics", no. 14. pp. 237–273.

Gierer A., Meinhardt H., *A theory of biological pattern formation*, "Kybernetik" 1972, no. 12, pp. 30–39.

Gros C., *Complex and Adaptive Dynamical Systems: A Primer*, Springer: Complexity 2010.

Kauffman S.A., *The Origins of Order: Self-Organization and Selection in Evolution* (OUP, USA) 1993.

Levitt S., Dubner S.J., *Freakonomics: A Rogue Economist Explores the Hidden Side of Everything*, William Morrow, HarperCollins 2005.

Nicolis G., Nicolis C., *Foundations of Complex Systems: Emergence, Information and Predication* (2nd ed.), World Scientific 2012.

Raine D.J., *Decoherence of the cluttered quantum vacuum*, [in:] *The Renaissance of General Relativity and Cosmology*, eds. G. Ellis, A. Lanza, J. Miller, Cambridge 1993, pp. 300–313.

Raine D., Symons S., *Emergence of coding in a prebiotic ecology*, JBPC 2009, no. 9, pp. 93–99.

Rees M., *Just Six Numbers: The Deep Forces That Shape the Universe*, Phoenix 2001.

Schelling T.C., "Journal of Mathematical sociology" 1971, no. 1, pp. 143–186.

Shnerb N.M., Louzoun Y., Bettelheim E., Solomon S., *The importance of being discrete: Life always wins on the surface*, PNAS 2000, no. 97 (19), pp. 10322–10324; www.pnas.org/cgi/doi/10.1073/pnas.180263697.

Uzan J-P., *Varying Constants, Gravitation and Cosmology*, *Living Rev. Relativity* 14, 2, 2011, http://www.livingreviews.org/lrr-2011-2012.

Wilczek F., *Fundamental constants*, [in:] *Visions of Discovery. New Light on Physics, Cosmology, and Consciousness*, eds. R.Y. Chiao et al., Cambridge 2010, pp. 75–104.

Wilensky U., *NetLogo Ising model*, http://ccl.northwestern.edu/netlogo/models/Ising. Center for Connected Learning and Computer-Based Modeling, Northwestern University, Evanston, IL 2003.

Zurek W.H., *Reduction of the wavepacket: how long does it take?* [in:] *Frontiers of Nonequilibrium Statistical Physics*, eds. G.T. Moore, M.O. Scully, Plenum Press, New York 1986, pp. 145–151.

Part II

Causality and the Structure of the Universe

Marek Kuś
Polish Academy of Sciences, Warszawa

The Uncertain Future and the Ambiguous Past in Classical, Quantum and General Non-signaling Settings

Discussion of the causal universe would remain incomplete without even a brief presentation of the problems of causal relations and causal explanations specific to quantum mechanics. The purpose of my contribution is to give a very short exposition of the topic or, better to say, some aspects of it as exhibited by very recent developments in the theory of quantum correlations.

With only slight exaggeration one can say that physics is in fact the art of determining correlations between various parts of physical reality. Indeed looking for universal laws consists in seeking correlations between some (say, initial) states of physical systems and their states in later instants of time, or dependence (i.e. again particular correlations) among different measurable quantities characterizing a physical system or its parts (subsystems).

To be more concrete, let us concentrate on a paradigmatic experimental setting in which we measure some properties of two spatially separated subsystems A and B. We use the term 'subsystems' since we want to treat A and B as parts of a composite system A + B . Depending on the situation results of measurements performed on both subsystems can be mutually dependent or not. For example, for isolated systems various physical quantities like energy, momentum or angular momentum remain constant in time. Measurements of such a quantity performed in two subsystems will be correlated since the amount pertaining to one subsystem is determined by the total (conserved) amount and the amount measured in the second subsystem. In fact, classical physics offers only two slightly different explanations of observed correlations. Either the subsystems A and B 'have a common origin' – they are spatially isolated parts of system which decayed at some moment of time (Fig. 1), or A and B were subject to some interaction between them spreading with a subluminal velocity (Fig. 2) causing exchange of information.

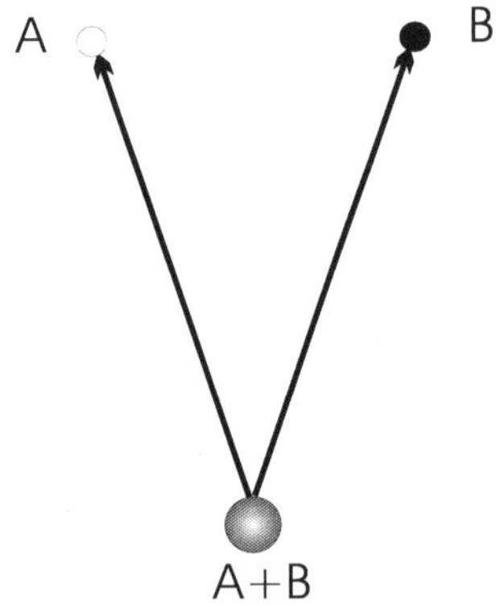

Figure 1

Two subsystems correlated due to a 'common origin'. A and B are parts of a large subsystem which decayed at some instant of time.

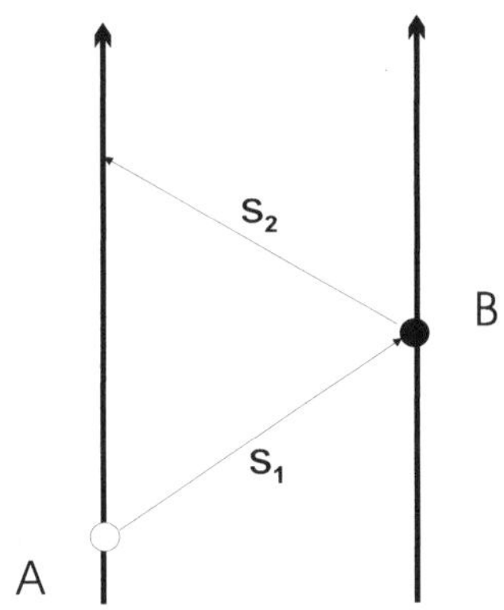

Figure 2

Two subsystems correlated due to a subluminal past exchange of information.

We can exclude the second possibility assuming that measurements performed on different subsystems are separated by time intervals short enough to assure that within them no information can be exchanged between different subsystems ('no communication during experiment'). Let us thus consider a general situation depicted in Fig. 3, where escaping parts $A_1,...,A_N$ of a defragmenting system cease to interact after some time t, and after this time we start our measurements observing the 'no communication' condition. In both classical and quantum settings we can position our experimental apparata at will, for example we can measure an arbitrary component of the angular momentum for each subsystem (in Fig. 3 it is symbolically depicted by differently oriented coordinate systems associated to each A_k).

Let us denote by x_1, x_2,...,x_N possible measurements (positions, momenta, spins, etc.), by a_1, a_2,...,a_N their possible outcomes and by

$$P\,(a_1, a_2,...,a_N | x_1, x_2,...,x_N) \tag{1}$$

the (conditional) probabilities of obtaining a_1, a_2,...,a_N if x_1, x_2,...,x_N were measured.

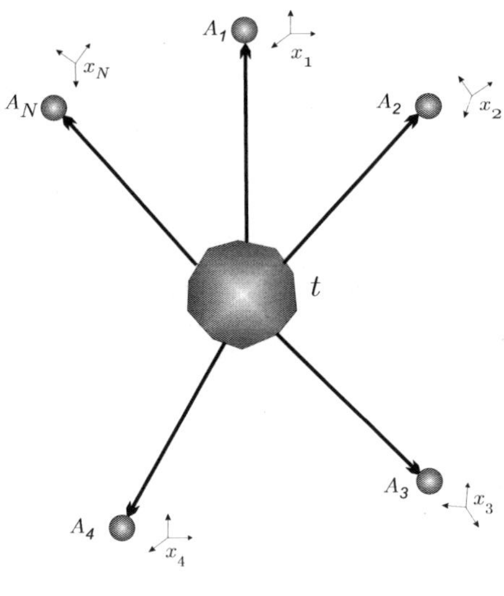

Figure 3

A system defragmented into N-parts mutually noninteracting after some time t.

In the simplest situation we have two subsystems under control of two distant laboratories. In each laboratory we can perform a measurement by positioning our measurement apparatus along some arbitrarily chosen direction x_i, $i = 1,2$ x_i , (Fig. 4). To fix the ideas we can think about measuring spin (angular momentum) components in directions perpendicular to the directions of motion of two particles resulting from some decay or, in the optical setting, polarizations of light emitted in two different directions by an atomic source.

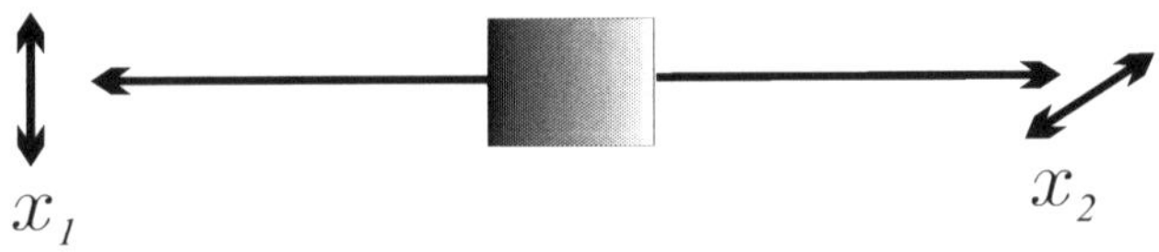

Figure 4

Measurements of spin components (polarizations) in spatially separated subsystems.

In the considered situation (remember that we assume that no exchange of information between subsystems is possible during the experiment) classical causality conditions concerning the choice of measurements x_1 and x_2 and their results a_1 and a_2 require that [1]:

• results of the measurements in one laboratory do not influence the probability of results in the second one

$$P(a_1 | a_2, x_1, x_2) = P(a_1 | x_1, x_2), \quad P(a_2 | a_1, x_1, x_2) = P(a_2 | x_1, x_2), \qquad (2)$$

• the choice of a measurement in one laboratory does not influence the probability of results in the second one

$$P(a_1 | x_1, x_2) = P(a_1 | x_1), \quad P(a_2 | x_1, x_2) = P(a_2 | x_2). \qquad (3)$$

Short calculations show that the above assumptions lead to an expected result: probabilities of outcomes depend only on local measurements, i.e.

$$P(a_1, a_2 \mid x_1, x_2) = P(a_1 \mid x_1)P(a_2 \mid x_2). \tag{4}$$

Indeed [1], from the definition of the conditional probability,

$$P(a_1, a_2 \mid x_1, x_2) = P(a_1 \mid a_2, x_1, x_2)P(a_2 \mid x_1, x_2), \tag{5}$$

and the causality conditions (2) and (3) we obtain

$$P(a_1, a_2 \mid x_1, x_2) = P(a_1 \mid x_1, x_2)P(a_2 \mid x_1, x_2) = P(a_1 \mid x_1)P(a_2 \mid x_2). \tag{6}$$

The ‘locality condition’ (4) means that results of measurements in each laboratory cannot be influenced by what happens in the second one or, in other words, results in each laboratory depend only on local arrangements. Such correlations admit a ‘hidden variables’ description [2],

$$P(a_1, a_2 \mid x_1, x_2) = \sum_\lambda p(\lambda) f_1(a_1 \mid x_1, \lambda) f_2(a_2 \mid x_2, \lambda). \tag{7}$$

Here $p(\lambda)$ is a probability distribution of (uncontrolled in the experiment) hidden variables λ. For each values of λ the ‘true state’ of the system is uniquely determined and so are outcomes a_i of any conceivable experiments x_i. In this way we achieve a fully classical description of measurement exactly as in classical statistical physics, where microscopic states are not controlled and the results of macroscopic experiments are obtained by averaging over all admissible microscopic configurations.

Local theories, admitting only local correlations (7) put some special limits on possible correlations among results of measurements called Bell (or Bell-like) inequalities [3]. One of the most known examples is the Clauser-Horne-Shimony-Holt (CHSH) inequality [4] applicable in the configuration shown in Fig. 4. The possible outcomes a_i can take only two values, say ± 1. We can think about measuring whether a single photon passes through a polarizer aligned with the direction x_i or determining the spin component along x_i of a spin-$\frac{1}{2}$ particle. In both cases the possible outcomes are dichotomous: a photon passes the polarizer ($a_i = 1$) or not ($a_i = -1$), the spin component s_{x_i} equals $\frac{\hbar}{2}$ which corresponds to, say, $a_i = 1$ or $s_{x_i} = -\frac{\hbar}{2}$ for which we take $a_i = -1$. Let us now consider a particular correlation,

$$\begin{aligned} E(x_1, x_2) &= P(1,1 \mid x_1, x_2) + P(-1,-1 \mid x_1, x_2) \\ &\quad - P(1,-1 \mid x_1, x_2) - P(-1,1 \mid x_1, x_2) \end{aligned} \tag{8}$$

or, in other words the difference between the probabilities of the same and opposite outcomes a_1 and a_2. Due to the fact that $a_i = \pm 1$ the correlation $E(x_1, x_2)$ equals simply the average value of the product $a_1 a_2$ which is equal to one if $a_1 = a_2$ and to minus one if $a_1 = -a_2$,

$$E(x_1, x_2) = \langle a_1 a_2 \rangle \tag{9}$$

Obviously the average $\langle a_1 a_2 \rangle$ depends, in general, on the actual alignments x_1, x_2 of the measuring apparata. By changing the alignments we can produce

$$S(x_1, x_1, x_2, x_2) = E(x_1, x_2) + E(x_1, x_2) + E(x_1, x_2) - E(x_1, x_2) \tag{10}$$

Using (9) we get

$$S := \langle a_1 a_2 + a_1 a_2 + a_1 a_2 - a_1 a_2 \rangle = \langle (a_1 + a_1) a_2 + (a_1 - a_1) a_2 \rangle = \langle \pm 2 \rangle \tag{11}$$

To see the last equality observe that one of the terms $(a_1 \pm a_2)$ must be equal 0 and then the other one amounts to 2. We thus obtain for the average the CHSH inequality

$$|S| \leq 2 \tag{12}$$

Quantum mechanics does not obey limitations imposed by the CSHS inequality. Quantum mechanically the correlation $E(x_1, x_2)$ has to be calculated as an quantum expectation value of the observables corresponding to the performed measurements. In the case of spin-$\frac{1}{2}$ particles it reads

$$E(x_1, x_2) = E_\psi(x_1, x_2) = \langle \psi \mid \vec{x}_1 \cdot \vec{\sigma})(\vec{x}_2 \cdot \vec{\sigma}) \mid \psi \rangle. \tag{13}$$

where $\mid \psi \rangle$ is the quantum state of the system, $\vec{\sigma}$ is the vector of the Pauli spin operators $\vec{\sigma} = (\sigma_1, \sigma_2, \sigma_3)$, and $\vec{x}$ are unit vectors in the directions along which components of spin are measured in a given experiment. If the state $\mid \psi \rangle$ is given as

$$\mid \psi \rangle = \frac{1}{\sqrt{2}}(\mid 1, 1 \rangle - \mid -1, -1 \rangle), \tag{14}$$

where $\mid 1, 1 \rangle$ and $\mid -1, -1 \rangle$ are the states of two spins being parallel and having, respectively, values $\frac{\hbar}{2}$ and $-\frac{\hbar}{2}$, then $E(x_1, x_2)$ reduces to

$$E(x_1, x_2) = -\vec{x}_1 \cdot \vec{x}_2 = -\cos(\alpha), \tag{15}$$

where α is the angle between the directions of measurements $\vec{x}_1$ and $\vec{x}_2$. Choosing the directions x_1, x_2, x_1' and x_2' as in Fig. 5 and calculating $S(x_1, x_1', x_2, x_2')$ from Equation (10) using (15) we obtain

$$|S| = 2\sqrt{2} \tag{16}$$

in a clear violation of the CHSH inequality (12). It was shown experimentally in an experiments with polarized photons that a corresponding inequality is indeed violated [5].

States like (14) are called entangled and do not appear in classical theory in which superpositions of distinct states (in this case $\mid 1, 1 \rangle$ and $\mid -1, -1 \rangle$) lack sense, but are admissible by quantum mechanics. We have thus to conclude that the causality principles (2) and (3) leading to locality are too strong. Quantum mechanics is non-local and does not admit a hidden variables model. In fact what we should demand is simply to exclude any information exchange between distant laboratories. Randomly chosen measurements x_i should give equally random

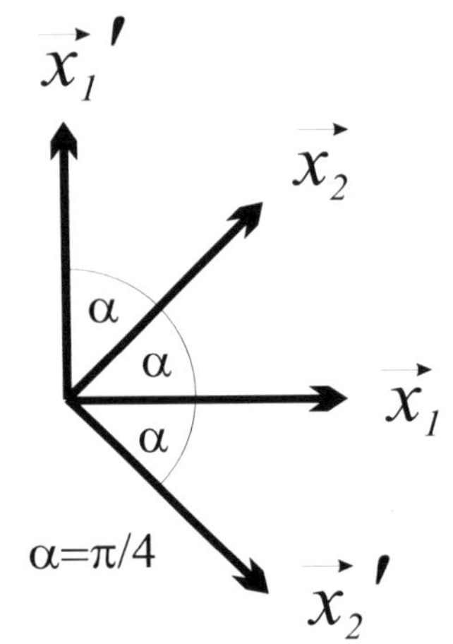

Figure 5

Configuration of polarizers in the Bell-like experiment.

outcomes a_i. This is guaranteed by the so called 'non-signalling' causality principle which in the two-particle setting reads

$$\sum_{a_1} P(a_1, a_2 \mid x_1, x_2) = P(a_2 \mid x_2) \tag{17}$$

$$\sum_{a_2} P(a_1, a_2 \mid x_1, x_2) = P(a_1 \mid x_1). \tag{18}$$

Both classical and quantum physics conform to this general non-signaling causality principle: the results of an experiment in a well localized laboratory cannot instantaneously influence physical states of distant systems. Interestingly, there are non-signalling theories which exhibit correlations stronger than is admissible by quantum mechanics. We have shown that all local theories (like classical statistical physics) impose bounds on S (10) in the form of the CHSH inequality (12). It was shown by Tsirelson [6] that there exits a corresponding boundary enforced by quantum mechanics

$$|S| \leq 2\sqrt{2}, \tag{19}$$

incidentally saturated by the setting presented above, cf. (16). Observe that without any additional knowledge of the nature of the correlations $E(x_1, x_2)$ except that they are bounded in their absolute value by 1 (as it was the case in both classical and quantum cases) we may hope to have $|S| = 4$ in (10) taking $E(x_1, x_2) = E(x_1, x_2^{'}) = E(x_1^{'}, x_2) = 1$ and $E(x_1^{'}, x_2^{'}) = -1$. In [7] Popescu and Rohrlich showed that this is possible within a non-signalling theory. In their model both, the measurement parameters x_i and the measurement outcomes take dichotomous values, $x_i = \pm 1$, $a_i = \pm 1$, and the connection between the measurements and their outcomes are given by

$$1 - \frac{a_1 + a_2}{2} = \frac{1}{4}(1 - x_1)(1 - x_2) \bmod 2 \tag{20}$$

It is easy to check that indeed the model is non-signalling, i.e. (17) and (18) are fulfilled, whereas $|S| = 4$. The question of how to restrict the general non-signalling causality principle in order to exclude theories admitting such 'superquantum', correlations is a challenging one [8].

I would like to complete my contribution by explaining its title by formulating a problem. Observe that the possibilities afforded by classical and quantum physics for acquiring knowledge about our past and future are strikingly different. In the classical setting, uncertainty about future events had an epistemological character – an event is random for us due to our lack of knowledge about all of its causes. Statistical physics at the turn of the 19th and 20th centuries, adopted (even if tacitly) this clearly Democtritean attitude – our lack of knowledge forces us to speak about phenomena in terms of averages and random correlations, but on the fundamental level the deterministic laws govern the motion of practically non-observable ('hidden') atoms or molecules unambiguously. In other words, classical physics does not leave any room for intrinsically random events. A purportedly random process is in fact completely determined and its course can be predicted with the desired accuracy once we improve our measuring instruments and the computational power of computers.

Although in the course of the development of quantum theory it became clear that it provides only a probabilistic description of reality, initially there were no reasons to switch to the Epicurean attitude stressing ontological sources of odds and chances – the deterministic motion of atoms is interrupted, without a cause, by 'swerves' which give rise to indeterminacy on higher levels. Quantum mechanics seemed to be another incomplete theory involving deterministically evolving hidden variables over which we have to average due to our inability to determine the exact values of them. Exhibiting nonlocal correlations violating Bell inequalities in an experiment gave proof of a nondeterministic nature of quantum mechanical reality and its intrinsic, ontological randomness.

It is interesting that 'treatment' of the past in both theories also exhibits significant differences [9]. In classical physics the past is, in principle, accessible with an accuracy limited only by our knowledge of all relevant circumstances and available computing capacity. It is not so in quantum mechanics. Let us consider a paradigmatic 'self-interference' experiment sketched in Figures 6 and 7. An incoming light beam is split by half-transmitting mirrors (depicted as the grey rectangles) into two beams and reflected by mirrors M_1 and M_2. In the first situation, one of the possible paths is blocked. Due to interference effects photons, after passing through the second half-transmitting mirror are observed with equal probability along two different directions of propagation. In this case, the path of the photon within the whole experimental apparatus is unambiguous. If, as in Fig. 8, both paths are free from obstacles we will observe photons propagating only in one direction after passing through the second half-transmitting mirror. Here any question concerning the path taken by a photon before emerging from the apparatus is senseless – the past of the photon is intrinsically ambiguous.

Now it is the proper time for the announced problem. Classical and quantum mechanics offer different mechanisms for making the past undetermined and the future uncertain – how do these ambiguities appear in other non-signaling theories? Does an answer (which

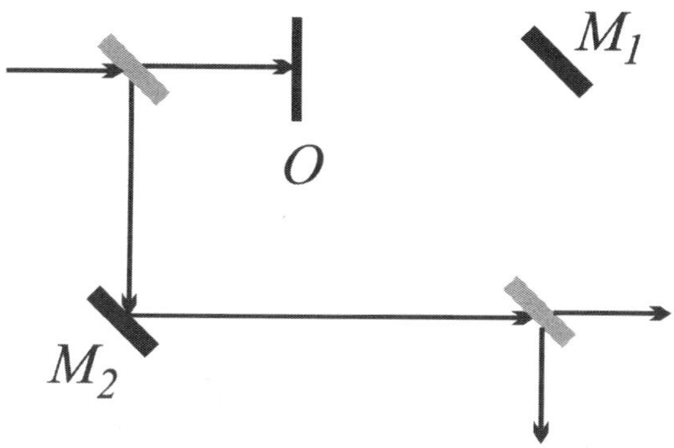

Figure 6

Two-way interference experiment with single photons. One of two possible path is blocked – there is no ambiguity in answering the question which path was traversed.

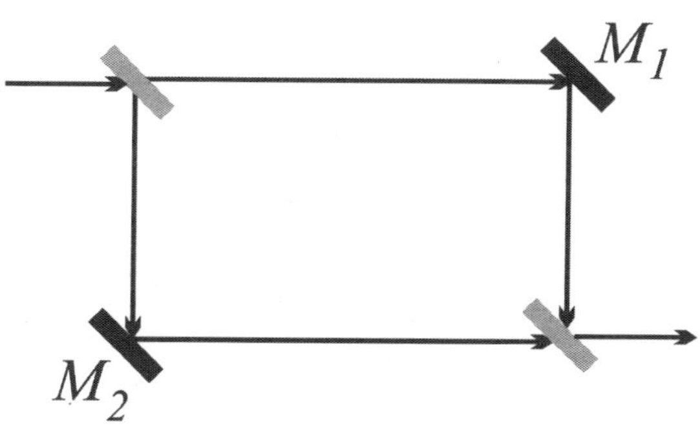

Figure 7

A photon can choose with equal probability both path determined by reflections from the mirrors M_1 or M_2. Due to interference effects only one output path will be observed. There is no method to determine which way was actually completed.

I do not know) to this question provide a way to decide why quantum mechanics is distinguished among all non-signalling theories by the fact that, as we believe, it offers an adaequate and correct description of reality?

References

[1] Ghirardi G.C., Romano R., "Int. J. Mod. Phys." 2013, no. B 27, p. 1245011.
[2] Masanes Ll., Acin A., Gisin N., "Phys. Rev." 2006, no. A. 73, p. 012112.
[3] Bell, J.S., "Physics" 1964, no. 1, p. 195.
[4] Clauser J.F., Horne M.A., Shimony A., Holt R.A., "Phys. Rev. Lett." 1969, no. 23, p. 880.
[5] Aspect A., Grangier P., Roger G., "Phys. Rev. Lett." 1981, no. 47, p. 460.
[6] Tsirelson B.S., "Lett. Math. Phys." 1980, no. 4, p. 93.
[7] Popescu S., Rohrlich D., "Found. Phys." 1994, no. 24, p. 379.
[8] Pawłowski M., Paterek T., Kaszlikowski D., Scarani V., Winter A., Żukowski M., "Nature" 2009, no. 461, p. 1101.
[9] Bennett C.H., http://www.hri.res.in/: iwqi12/iwqi12_talks/CH_Bennett-Triveni.pdf.

Julian Barbour
College Farm, South Newington, Banbury

Reductionist Doubts[1]

1. Reductionism's Strengths and Weaknesses

> Nature does not begin with elements as we are obliged to begin with them. It is certainly fortunate that we can, from time to time, turn aside our eyes from the overpowering unity of the All and allow them to rest on individual details. But we should not omit ultimately to complete and correct our views by a thorough consideration of the things which for the time being we left out of account (Ernst Mach 1883).

To get an idea where reductionism's strong and weak points lie, let's go to its source in Newton's worldview: utterly simple laws that govern the motions of bodies in space and time.

---

[1] This paper is my entry in the Fourth Essay Competition of the Foundational Questions Institute (fqxi.org), the topic of which was 'Questioning the Foundations: Which of Our Basic Physical Assumptions Are Wrong?' In my essay, I consider some of the implications of the approach to dynamics called shape dynamics that I discussed at the 16th Kraków Methodological Conference Causal Explanations in Physics and Cosmology. Since shape dynamics is adequately presented in references [2, 6, 5, 7, 8] of the present paper, there seemed to me little point in simply reproducing already-published material. Instead, it seemed to me more profitable to consider some of the implications of shape dynamics for our notions of the causal interconnections of the universe, as I do in this essay. I am grateful to Michael Heller and the other organizers of the Kraków conference for the invitation to speak at it and to include the present essay in the proceedings. I am also very grateful to the Foundational Questions Institute for research grants during the last four years (RFP2-08-05 and 2010 Time and Foundations RFP).

The law of inertia is the most basic: a force-free body moves uniformly in a straight line forever. Things are almost as simple if bodies interact. When far apart, they move with near perfect inertial motions, but when closer they can, through forces like gravitation, begin to change each other's motions. If body A has a greater mass than body B, then A affects B more than B does A. With allowance for the masses, action and reaction are equal.

The behaviour of a large system of many bodies, in principle the whole universe, can, according to Newton, be entirely explained by the inherent tendency of each body to move inertially modified merely by the forces exerted by all the other bodies in the universe. Just as reductionism proposes, the whole truly is the sum of its parts. Or is it?

The weak spot in Newton's scheme is the very thing that makes it reductionist. The position and motion of individual bodies such as atoms, the simple entities, are defined relative to invisible space and time, the framework in which simple laws can be formulated. Mach [1] and others argued that in reality the position of any one object is defined relative to every other object in the universe. That is obviously a far more complicated approach and it is clearly holistic. Mach argued that nevertheless it could still reproduce all of Newton's successes because force-free bodies are observed to move rectilinearly relative to the stars, which suggests that, in their totality, they exert a powerful causal effect on individual objects. Newton could have mistaken this empirical fact as evidence for what he called absolute space.

It is easy to see how this could have led to seriously wrong ideas about the universe. Barely over 100 years ago, most scientists thought we lived in an 'island universe' of a few million stars (our Galaxy) with nothing but space outside it. According to Newton's laws, such an island universe could exist and have angular momentum, L, about an axis, making it oblate like the rapidly rotating Jupiter. This was to be expected.

However, a simple implementation of Mach's ideas [2] rules it out. A Machian island universe must have angular momentum *exactly equal to zero*: $L = 0$, even though subsystems of bodies in small regions of the Galaxy can behave as Newton's laws predict and have $L \neq 0$. It is merely necessary for the values of L for all the subsystems to add up to zero. Seeing Newton's laws confirmed in the solar system, which does have $L \neq 0$, astronomers had no reason to question any of the predictions of Newtonian theory, which in no way is capable of forcing an island universe to have $L = 0$.

But suppose astronomers had, on Machian grounds, been convinced around 1900 that the universe must have $L = 0$ and had found this was not so for the Galaxy. Far from giving up the Machian theory, they would have *predicted* that our Galaxy cannot be the entire universe. They would have predicted a universe with more matter whose angular momentum exactly balances the Galaxy's.

How does the universe we can now observe meet this expectation? Extraordinarily well. It is full of galaxies and clusters of galaxies. Each and every one manifestly possesses some angular momentum, but there is no indication that the individual angular momenta add up to give a significant non-zero angular momentum of the universe as a whole.

In fact, my example is too simplistic, being based on Newtonian theory and not its superior successor in the form of Einstein's general relativity. But I think it makes my point. A reductionist standpoint may be very misleading.

2. Einstein, Mach, and General Relativity

Mach's idea that Newton's invisible absolute space could be replaced by an effect of the entire universe had a deep influence on Einstein, who called the idea *Mach's principle* [3]. It was the single greatest stimulus to the creation of his wonderful theory of gravity, general relativity (GR).

If GR implements Mach's principle, reductionism will be challenged. Simple parts and their interactions will not determine the whole; the whole will determine the way the parts behave. In fact, we shall see that the whole to a considerable degree determines what we perceive as parts.

So does GR implement Mach's principle? This has been a matter of controversy ever since Einstein created GR, mainly because he set about implementing Mach's principle in an indirect manner that left its status in GR obscure [4]. In fact, his definition of Mach's principle [3] was inconsistent [4], p. 93. I have given what I believe is the correct definition of Mach's principle [5] and argued that if the universe is closed up in three dimensions like the earth's surface in two then GR does implement Mach's principle [5]. If the universe is spatially infinite, the answer is equivocal. It is Machian however far you can imagine, but infinity is unreachable, and one can never establish a complete sense in which the whole determines the part.

Section 1 showed how reductionism can mislead if we conceive the universe as a mere sum of its parts, but it did not suggest any need to change the conception of the parts, the individual stars, but only the way they interact. The Machian interpretation of general relativity is much more sophisticated and changes radically the way we conceive the parts and not just the way they interact. In this section I show how the standard representation of GR presents the universe as a sum of parts. In Sec. 3, I give the alternative Machian interpretation, in which GR appears much more holistic. In the final section, I discuss possible implications of a holistic quantum view of the universe.

By parts in GR I mean little (infinitesimal) regions of spacetime. Imagine a two-dimensional surface that, at each point, is smooth and curved in accordance with a definite local law: measurements of the curvature in the immediate vicinity of every point on the surface would confirm the law holds at that point. Any surface for which this is true may exist. This is a reductionist situation: the parts (infinitesimal elements of surface) and the local law they satisfy determine what wholes (extended two-dimensional surfaces) can exist.

In general relativity, in the simplest situation in which no matter is present, the infinitesimal elements have not two but four dimensions: one of time and three of space. Otherwise the situation is very similar to what I described. Infinitesimal regions of spacetime satisfy Einstein's famous field equation $R_{\mu\nu} = 0$, and his theory permits any spacetime in which this local law is satisfied everywhere. Presented in these terms, it is a great triumph of reductionism; the predictions of GR have been very well confirmed for subsystems of the universe (the solar system and binary pulsars) and are not in conflict with cosmological observations. But in fact these present numerous puzzles whose solution may well call for a truly holistic approach that restricts the solutions GR allows and puts them in a different perspective. I now turn to that.

3. Gravity, Angles, and Distances

In Sec. 2, I identified the 'parts' in GR as infinitesimal pieces of spacetime. Their structure is determined by the fundamental quantity in GR: the metric tensor $g_{\mu\nu}$. Being symmetric ($g_{\mu\nu} = g_{\nu\mu}$) it has ten independent components, corresponding to the four values the indices μ and ν can each take: 0 (for the time direction) and 1,2,3 for the three spatial directions. Of the ten components, four merely reflect how the coordinate system has been chosen; only six count. One of them determines the four-dimensional volume, or *scale*, of the piece of spacetime, the others the *angles* between curves that meet in it. These are angles between directions in space and also between the time direction and a spatial direction.

Now Mach's attitude to physics was not only holistic but also that all theoretical concepts should stay as close as possible to directly observable quantities. Using grand philosophical terms, the gap between epistemology – what can be observed – and ontology – what is assumed to exist – should be as small as possible. Ideally, there should be no gap at all. From this perspective, Einstein's ontology as expressed through the metric tensor $g_{\mu\nu}$ can be questioned on two grounds.

First, time is taken to be just as real as space. But Mach argued that time cannot be a primary notion; it must be derived from motion. In the first of these fqxi essay competitions, I showed how Newton's notion of an independent absolute time can be derived from motion. In Einstein's theory, Newton's global absolute and nondynamical time is replaced by local proper time, which interacts with space. This is a major change, but time, like space, is still taken to be a primary ontological concept, not something to be derived from more basic concepts.

Second, there is a big difference between, on one hand, quantities with scale, like lengths and volumes, and, on the other, angles: the former cannot be specified without the human choice of some unit – say metres or yards. That is arbitrary. In contrast, an angle can be specified in dimensionless terms as a fraction of a radian, which itself is a dimensionless quantity.

Let me spell this out to underline the Machian aspect. Suppose yourself at the centre of a circle of radius r looking at two points on the circumference separated by r. The angle that you see between them is by definition one radian. Key here is that the angle is the same whatever the radius r. This brings out the difference between angles and lengths.

Now imagine yourself on a clear night at high altitude in a desert. What do you see? Thousands of stars studding the black sky, all with definite angles between them. It's beautiful. The scientific point is that you directly see the angles between the stars. But you cannot tell their distances[2]. There are two possible reactions to the difference between dimensionful and inherently dimensionless quantities like angles. The standard one in science is to say that it is not a big deal; one must simply express everything in terms of ratios. Astronomers, for example,

---

2 Even when distances are determined by parallax, it is changes of observed angles that determine the distances, which are themselves ratios of the trigonometric base length.

express interplanetary distances as ratios, taking the semi-major axis of the Earth's orbit as unit. Moreover, these distances have *dynamical* effect: Newton's gravitational force is inversely proportional to the square of the distance.

The alternative attitude to dimensionful quantities is to deny them any fundamental role. Is this feasible? I think so. Note first that, despite being defined through ratios of distances, angles can be truly said to belong to a point. The radius of the circle taken to define a radian can always be taken as small as one likes. In contrast, a distance is necessarily associated with two points, the ends of the considered interval. In the standard formulation of GR, angles and distances both have ontological – dynamical – status. Distances do because the curvature of spacetime plays a central role in GR, and curvature is defined by comparisons over distances even though they can be taken arbitrarily small[3]. Curvature, and with it distance, is so deeply embedded in the conceptual and technical foundations of GR it would seem only a madman would try to deny it a deep dynamical significance. General relativity without curvature seems like Hamlet without the Prince of Denmark. Remarkably, a new formulation of GR, called *shape dynamics* [6, 7, 8], suggests that, in a well defined sense, it is only angles that count. Because it shows how reductionism could be misleading us, I will sketch how the significance of angles emerged.

Dirac [9] and Arnowitt, Deser, and Misner (ADM) [10] took the first step over 50 years ago by replacing Einstein's vision of spacetime as a four-dimensional 'block' universe by a picture in which a three-dimensional entity evolves, building up spacetime in the process. This is known as the ADM formalism. The hope, still held by many, is that the Dirac–ADM approach will one day lead, as its creators intended, to a quantum theory of gravity.

I need to say what the Dirac–ADM three-dimensional 'entity' is. It is a Riemannian three-geometry (named after the great mathemati-

[3] Technically, curvature involves second derivatives of the metric tensor, whereas angles are defined by the undifferentiated metric tensor.

cian Riemann, who introduced the concept). A two-geometry is easy to visualize: it is like the two-dimensional curved surface of the Earth, which closes up on itself. A closed geometry is needed to model a universe as a whole. A closed three-geometry is much harder to imagine, but is mathematically possible. In the ADM formalism angles and infinitesimal distances are, dynamically, on an equal footing.

However, a hint that angles might have priority emerged from subsequent work related to the technically important ADM initial-value problem. There is no need for me to give details except to say that without its solution it is simply impossible to make any practical use of GR to establish, for example, what happens when two black holes coalesce. Among other things, this work, by James York [11] with input from his student Niall Ó Murchadha [12], established the amount of information that is needed to determine a physically sensible solution of GR.

The required information can be expressed in terms of a 3-geometry and the rate at which it is changing. As with spacetime, at each point of a 3-geometry there is encoded angle and scale information, 2 and 1 numbers respectively (compared with 5 + 1 for four-dimensional spacetime). The rates at which these are changing are also characterized by 2 + 1 numbers. York showed that the purely angle part of this information, i.e., 2 + 2 = 4 numbers at each point, is sufficient to solve the initial value problem and hence determine a complete spacetime that satisfies Einstein's equations.

This suggested that GR is a theory of the way angle data change and that distance data play no dynamical role. However, this possibility was never taken too seriously because York's technique seemed to violate a foundational principle of GR – that there is no distinguished definition of simultaneity in the universe. For this reason, York's work, which does single out a notion of simultaneity, has usually been regarded as a merely technical device, albeit very valuable, for solving problems but not settling issues of principle.

4. Shape Dynamics

To what extent does shape dynamics (SD), which I mentioned earlier and for which the above has been a preparation, change this?

Imagine a 'toy' model universe of just three particles in two dimensions. Picture them as dots on an infinite sheet of grid paper. Then two coordinates determine the position of each. The six coordinates together define a *Newtonian configuration*, the sheet playing the role of absolute space. But one can ignore the positions of the dots on the sheet and regard as real only the three distances between each particle pair. That defines a Machian *relative configuration*. Defined in its own intrinsic terms, it does not exist anywhere in space. The final step is to say only the shape of the triangle, defined by two of its angles, is real.

The progressive elimination of non-Machian data takes us from Newtonian configurations through relative configurations to *shapes*. Now comes a key fact. Mathematically it is vastly easier to describe change in terms of coordinates, at the level of Newtonian configurations. That is why Newton introduced absolute space – and with it reductionism. Because shapes are irreducibly holistic, it is much harder to work with them and achieve the gold standard of dynamical theory: the ability to determine the future from an initial state of the system. In Newton's dynamics, a configuration and its rate of change define an initial state. In shape dynamics, we want a shape and its rate of change to determine an initial state[4]. That's where the problem, the

[4] I need to say something about what rate of change means in shape dynamics. It is not with respect to any external time but relates to the expansion of the universe, which cosmology indicates is a reality. In the context of shape dynamics, this introduces a single overall scale ratio: one can say the universe is twice as large now as it was in an earlier epoch. This scale ratio, which is why shape dynamics is based on volume-preserving conformal transformations, has nothing to do with angles but provides a single global parameter that defines the rate of change of the angles. Thus, the angles do not depend on time but on the scale ratio. I do find the need for a scale ratio mysterious. Perhaps it has to be there to provide a substitute for time.

passage from reductionism to holism, lies. It is solved by *best matching*, which is explained in Fig. 1.

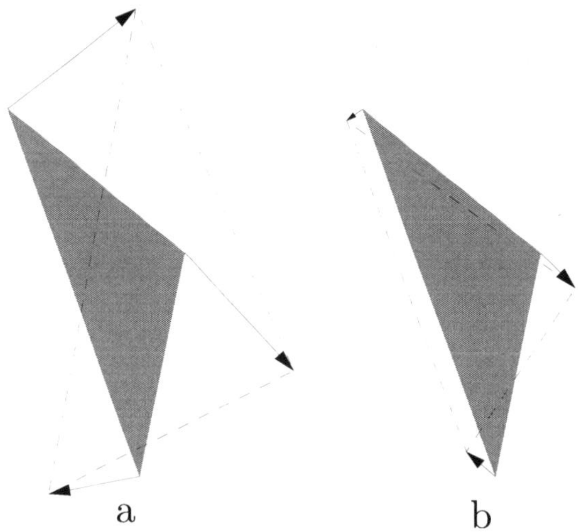

Figure 1

a) An arbitrary relative positioning of two triangles determines an initial Newtonian configuration (vertices of the grey triangle) and its rate of change (length of the arrows to the vertices of the dashed triangle). In the best-matched positioning b), shifting, rotating and scaling has brought the dashed triangle as close as possible to perfect overlap (congruence) with the grey one. An appropriate mathematical expression [5] measures the residual incongruence, the determination of which depends only on the shapes, not on their sizes or positions in space. The procedure is holistic, since the shapes alone determine the outcome, which is moreover expressed tractably through the coordinates of the vertices of the dashed triangle.

Once the shape-dynamic problem of defining initial states has been solved, it is easy to determine how shapes change according to a law that in no way relies on absolute duration, on position in an frame like grid paper, or on an external scale that measures size. The idea can be applied very generally, in particular to Riemannian geometries from which all information except that relating to angles between curves has been abstracted. Then we are dealing with *shapes of closed geometries*, just as York did, but now with some differences.

First, SD has led to an improved understanding of York's technique. This is a technical matter that I will relegate to an end note[5] for experts.

Second, and relevant for us, York did not solve the initial-value problem following any clearly formulated first principles but by exploring different mathematical techniques until he found one that worked, supremely well in fact. In contrast, SD derives not only Einstein's equations but also York's method for solving the initial-value problem on the basis of clearly formulated Machian first principles, of which there is a 'holy trinity': duration, position, and size must all be relative. The first two of these were foundational principles of Einstein's original derivation, but the third was not. In the shape-dynamic reformulation of GR, Einstein's relativity of simultaneity is traded for relativity of size.

The generalization of best matching to 'shapes of geometries' is direct but so far reaching it is worth outlining. One starts from 2 + 2 given pieces of angle information at each point and considers how one could add to them in all possible ways 1 + 1 pieces of 'trial' information that specify distances and their rates of change. The extra 1 + 1 pieces of information are then varied in all possible ways until best matching is achieved. The mathematics is sophisticated but conceptually fully analogous to adjusting the dashed triangle to the best-matched positioning in Fig. 1. As in that case, there is always just one unique way, for given 2 + 2 angle information, in which the best-matching condition can be satisfied. One then has an initial state that determines the future uniquely[6].

---

[5] York based his technique, for a spatially closed universe, on full conformal transformations (which change local sizes but leave the angle information unchanged), whereas SD is based on volume-preserving conformal transformations. This is a tiny restriction, but it explains a bizarre feature of York's method that seemed completely *ad hoc*. I am referring to the scaling law that York adopted for the trace part of the momentum in the Lichnerowicz-York equation, which seemed incompatible with the law adopted for the trace-free part. SD provides a simple explanation for the law [6].

[6] My characterization of shape dynamics emphasizes its conceptual aspects. My collaborators Henrique Gomes, Sean Gryb, and Tim Koslowski [8] also use the name shape dynamics for a specific gauge-theoretical implementation of the underlying ideas in a form that is likely to be best suited for technical developments.

Prior to the best matching, there is *nothing* in the specification of the initial 2 + 2 pieces of angle information at each point of space that determines the distance between nearby points, the rate at which time flows at each space point, or the path of an inertially moving body. What I find really remarkable is that best matching applied to the distance-free 2 + 2 pure angle information leads to complete determination of distances, rates of time flow, and inertial motion everywhere[7]. It is a triumph of Mach's intuition, going far beyond his original tentative suggestions.

Best matching is profoundly holistic. The procedure based on it that determines how the extra 1 + 1 pieces of distance information are added at each point has to take into account the angle information *everywhere* – at all points. What is fixed *here* is determined by everything that is *there*[8]. There is a remarkable delicate interdependence of everything.

You may well ask: has anything been achieved, or is SD merely a different way of interpreting GR? The real test will be whether SD helps in the creation of a quantum theory of gravity. I shall consider that in Sec. 5. It is however noteworthy that GR allows a great number of solutions that seem physically most implausible; this has long been a concern. The ADM formalism, if regarded as the true form of the theory, already considerably restricted the allowed solutions. Shape dynamics goes significantly further. It requires the universe to be compatible with generation from York initial data and to be spatially closed. This what the truly holistic approach requires. In this respect, it runs counter to much current thinking in cosmology. It is always good for a theory to make bold predictions and 'live dangerously'.

---

[7] Soon after its completion, York's work suggested to Isenberg and Wheeler [13] a formulation of Mach's principle in somewhat similar terms but without a clear shape--dynamic underpinning.

[8] Best matching in the case of 'shapes of geometries' involves the solution of elliptic partial differential equations. As York and Ó Murchadha showed, these have very good existence and uniqueness properties. In the case of spatially closed universes, there are no boundary conditions to be specified; everything is determined intrinsically and holistically.

The main aim of this section has been to show that a reductionist approach can give a seriously misleading intuition about 'what the world is made of' and how it 'works'. In certain key respects, the standard representation of GR remains close to Newton's world view. Time and space are fused together, but locally their ontological nature and inertial motion in them are barely changed. Distance in space and duration in time are both real and foundational. Shape dynamics suggests that only angles are real[9].

5. Quantum Implications

Consider the archetypal quantum two-slit experiment in which, in Dirac's famous words, 'each photon interferes with itself'. This implies a particle that somehow splits in two and manages to 'go through' both slits at once. This obviously stretches the imagination.

Could a reductionist mindset be misleading us in trying to interpret quantum experiments in terms of particles moving in space and time? It's a very Newtonian picture. We see the Moon but never a photon. All we see is equipment and an event that we attribute to a photon. We see a photon-generating laser, a filter to ensure that only one photon at a time reaches the slits, and then the screen on which spots appear one after another.

[9] I should add two caveats here. First, the discussion so far has ignored matter. That can be treated by York's method and hence added to shape dynamics without difficulty. Indeed it adds conceptual clarity to the picture. This is because matter fields define curves in space. The angles between any two curves are real (ontological), but distances along the curves are not; they are gauge. Second, all relativists would agree that a spacetime is determined by the specification of 2 + 2 pieces of information at each point in a three-geometry. However, a majority would probably deny that these pieces of information must be exclusively and exhaustively angle data; they could be a mixture of angle and distance data. If relativity of simultaneity is taken to be sacrosanct, that is undeniable but it leaves one with a very indeterminate situation. What, if any, mixture of angle and distance data is correct? Three arguments speak for pure angle data: angles are conceptually more fundamental, the choice is clean (no indeterminate mixture), no other general and robust method apart from York's pure-angle method has been found to solve the initial-value problem.

The observed events are certainly discrete, but are we right to try to interpret them in terms of point-like particles? We do so because the experiment results in those spots on the screen or some other localized responses in apparatus. Of course, Bohr above all emphasized the complementary nature of quantum phenomena: particle-like or wavelike features are manifested depending on the macroscopic arrangement of the experiment. Change the arrangement and complementary aspects of reality are manifested. One can measure either momentum or position of a particle but never both at once.

This led Bohr to insist on the impossibility of formulating any meaningful statements about quantum phenomena without the framework provided by the classical, non-quantum world of macroscopic instruments. In this view quantum theory is not about 'what is' but about 'what we can say'.

But perhaps quantum theory *is* about 'what is' and the problem is that reductionism has led us to the wrong picture of 'what is'. Instead of thinking of particles in space and time, we should perhaps be thinking in terms of complete shapes of the universe. Then a photon would not be a particle that moves from one place to another but a law telling us how one shape of the universe becomes another. After all, that is what happens in the two-slit experiment. A spot appears on a screen that was not there before. In the first place, this tells us the shape of the universe has changed. Of course, it is only one change among an immense multitude, but it is a part of the totality of change. We may be doing violence to the universe – Mach's overpowering unity of the All – by supposing individual events are brought about by individual things, by parts. Hard as it may be, I think we need to conceptualize in terms of wholes.

The Machian interpretation of GR gives us hints of how this could change things. It suggests that what we observe locally has a part that is 'really here' – in the first place angle information – and a framework and basic laws that 'seem to be here' – local distances, times and inertial motion – but are in reality the effect of the complete universe.

In many ways, Bohr had a Machian mindset: the interpretation of quantum events depends crucially on the relationships that define the experimental layout. But where does that end and what determines it? Quantum entanglement measurements are now performed between the Canary Islands and the west coast of Africa and would be impossible without clocks that measure photon travel times. But if the Machian interpretation of GR is correct, the very distances over which travel times can be measured and even the times themselves are in a real sense created by the universe.

Thus, I think Bohr was on the right track but that one needs to go much further along it. Where will it take us? I don't know but will hazard a conjecture or two.

First, the quantum mechanics of the universe, like the Machian interpretation of GR, is holistic: it is about shapes. However, in contrast to shape dynamics, in which one shape follows another, in the quantum universe *all* shapes are present with different probabilities. But then whence comes our tremendously strong impression that time passes and we have a unique history?

In *The End of Time* [14], I attempted to answer this question with the notion of a *time capsule*: a very special configuration of the universe whose structure suggests that it is the outcome of a process that has taken place in time in accordance with definite laws. In [14] I still took distance to be real. The ideas that later led to shape dynamics were only briefly mentioned (in Box 3). I now distinguish carefully between relative configurations, in which both distances and angles are fundamental, and shapes defined by angles alone.

In [14] I conjectured a 'timeless quantum competition' between all possible configurations of the universe in which time capsules get the greatest quantum probabilities. We experience them as instants of time bearing every appearance of the transition from a definite past to an indefinite future. I justified this conjecture by the pronounced asymmetry of what I called *Platonia*, the collection of all possible relative configurations of the universe. But with distance removed from the foundations, we are left with shapes. Platonia is *shape space*.

In this view, the quantum universe is *doubly holistic*: the possible shapes of the universe are wholes, each a unity in itself, but there is also shape space, the higher whole formed by the totality of possible wholes. Shape space is the ultimate quantum arena[10]. That is my second conjecture.

Seen from a 'god's eye' perspective, the striking asymmetry of Platonia that I noted in [14] becomes even more pronounced and suggestive: in any space of shapes there is always one unique shape, appropriately called Alpha, that is *more uniform* than any other possible shape.

It interesting to note in this connection that the theory of cosmological inflation has some significant achievements to its credit but nevertheless must assume, as a hitherto unexplained initial condition, that the presently observed universe emerged from a state that was highly isotropic and homogeneous: in a word, very uniform. This suggests to me, as my third conjecture, that we should not be looking to explain the isotropy and homogeneity by a special initial condition but through the dominant status of Alpha in the universe's shape space. This is reflected in the fact that every shape has a degree of complexity that ranges from the most uniform possible to shapes of ever increasing complexity[11]. One never reaches Omega.

---

[10] In the ADM approach to quantum gravity, the wave function of the universe for vacuum gravity is defined on superspace, the space of Riemannian three-geometries. The problem with this arena is the supreme difficulty, unresolved despite over 50 years of attempts, of implementing the ADM Hamiltonian constraint. I suspect that the root of the problem is the indeterminate mixture of angle and distance information encoded in superspace, which, as noted in the main text, is unavoidable as long as relativity of simultaneity (refoliation invariance) is held to be sacrosanct. In shape dynamics, the wave function of the universe is defined on conformal superspace. This brings undoubted conceptual clarity, as York already pointed out, but severe problems still remain. The ADM Hamiltonian may be conceptually hybrid but it is at least local; the shape-dynamic Hamiltonian is non-local.

[11] Complexity is a surprisingly difficult notion to capture in a definition. However, in the case of the Newtonian N-body problem, there is a uniquely appropriate measure of the complexity of an N-body system that I call the scale-invariant gravitational potential V_{si}. It is the product of the ordinary Newtonian potential and the square root of the centre-of-mass moment of inertia $I_{cm} := (1/M)\sum_{i<j} m_i m_j r_{ij}^2$, where r_{ij} is the distance between particles i and j and M is the total mass. It is a function on shape space and takes its

If there is no Omega in shape space, there must still be an end to this essay. Let it be my final conjecture, which sharpens the one in [7]. Our experiential life is dominated by time and a great asymmetry: the transition from past to future, from birth to death. Science has so far failed to explain the asymmetry and has been forced to attibute it to a special – *highly uniform* – initial state of the universe. I suspect we need look no further than the structure of shape space. Its highly uniform Alpha, which holistic shape dynamics tells us must be deeply significant, seems to match perfectly what we take to be the uniform initial state of the universe.

Mathematics, the tool of theoretical physics, can only describe becoming through differences of structures that simply are – they exist in a timeless Platonic realm. Becoming is the ever mysterious moving reflection of being. We cannot explain the indubitable asymmetry of becoming that we experience unless being is asymmetric. It is.

maximum on the most uniform shape the system can have. For large N, the corresponding distributions of the particles are remarkably uniform. As the distributions become more clustered, V_{si} takes ever larger negative values. Given the close analogy between the Newtonian N-body problem when recast in Machian terms and GR, I anticipate the existence of a function on conformal superspace (the shape space for GR) that is analogous to V_{si}.

References

[1] Mach E., *Die Mechanik in ihrer Entwicklung Historisch-Kritisch Dargestellt*, 1883.
[2] Barbour J., Bertotti B., *Mach's principle and the structure of dynamical theories* (downloadable from platonia.com), Proceedings of the Royal Society London A 1982, no. 382, pp. 295–306.
[3] Einstein A., *Prinzipielles zur allgemeinen Relativitätstheorie*, "Annalen der Physik" 1918, no. 55, pp. 241–244.
[4] Barbour J., Pfister H.,, eds., *Mach's Principle: From Newton's Bucket to Quantum Gravity*, volume 6 of Einstein Studies. Birkhäuser, Boston 1995.
[5] Barbour J., *Mach's principle: A response to Mashhoon and Wesson's paper* arXiv: 1106.6036 (arXiv: 1108.3057), "Annalen der Physik" 2011.
[6] Anderson E., Barbour J., Foster B.Z., Kelleher B., Ó Murchadha N., *The physical gravitational degrees of freedom*, "Class. Quantum Grav." 2005, no. 22, pp. 1795–1802, gr-qc/0407104.
[7] Barbour J., *Shape dynamics: An introduction*, [in:] *Quantum Field Theory and Gravity. Proc. Conference at Regensburg 2010*, eds. F. Finster et al., Birkäuser 2012.
[8] Gomes H., Gryb S., Koslowski T., *Einstein gravity as a 3D conformally invariant theory* (arXiv:1010.2481), "Class. Quant. Grav." 2011, 28:045005.
[9] Dirac P.A.M., *Fixation of coordinates in the Hamiltonian theory of gravitation*, "Physical Review" 1959, no. 114, pp. 924–930.
[10] Arnowitt R., Deser S., Misner C.W., *The dynamics of general relativity*, [in:] *Gravitation: An Introduction to Current Research*, ed. L. Witten, Wiley, New York 1962, pp. 227–265.
[11] York J.W., *The role of conformal 3-geometry in the dynamics of gravitation*, "Phys. Rev. Letters" 1972, no. 28, pp. 1082–1085.
[12] Ó Murchadha N., York J., *Existence and uniqueness of solutions of the hamiltonian constraint of general relativity on compact manifolds*, "J. Math. Phys." 1973, no. 4, pp. 1551–1557.

[13] Isenberg J., Wheeler J.A., *Inertia here is fixed by mass–energy there in every W model universe*, [in:] eds. Pantaleo M., de Finis F., *Relativity, Quanta, and Cosmology in the Development of the Scientific Thought of Albert Einstein*, Vol. I, Johnson Reprint Corporation, New York 1979, pp. 267–293.
[14] Barbour J., *The End of Time*, Weidenfeld and Nicolson, London; Oxford University Press, New York 1999.
[15] Battye R., Gibbons G., Sutcliffe P., *Central configurations in three dimensions*, "Proc. Roy. Soc. Lond." 2003, no. A459, pp. 911–943.

Andrzej M. Sołtan
Nicolaus Copernicus Astronomical Center, Warszawa

X-ray Background and Cosmology

1. Introduction

For centuries cosmology was based more on philosophical beliefs rather than on the empirical facts. This relationship began to change after Newton's discovery of the law of universal gravitation. The next important step was made by application of Einstein's General Relativity to simple models of an isotropic and spatially homogeneous universe. The works of Friedman, Lemaître and Hubble have provn that modern physics and astrophysics embody adequate means to investigate the global properties of the real Universe. From that time, cosmology has been established definitely as a branch of natural science. Although some fundamental questions of a cosmological nature are still prescribed by philosophical assumptions, all the quantitative answers are obtained exclusively by empirical methods.

Due to obvious reasons we will not be able to examine all the Universe, if it is infinite. Hence, from a purely observational point, questions about the global characteristics of the Universe, such as the average energy density, the average chemical composition, or the average properties of objects populating the whole Universe, cannot be answered. To overcome this obstacle, one may *assume* that – for instance – the Universe is homogeneous above some a priori unknown spatial scale. Then it is a matter of observation to determine the extent of the area which appears to be representative for the whole observable volume of the Universe. As long as this measurement is not falsified by other observations, parameters determined for that area are considered to hold for the entire Universe.

Along this scheme, in the last hundred years many branches of astrophysics have contributed to the advancement of modern cosmology. Perhaps the most prominent example of the observation covering very limited volume of space which was extended to the entire Universe was the Hubble measurement of radial velocities of dozen or so galaxies situated in the immediate vicinity of our Galaxy. The Hubble law and universal 'escape of galaxies' have led to the hot Big Bang model, one of the most elegant scientific theories. The hot beginnings of the Universe have been confirmed observationally in the 1960's by the discovery of the cosmic microwave background (CMB) radiation. Investigation of the CMB in turn has allowed the further development of Big Bang theory.

Background radiation, of which the CMB is the best known component, plays a variety of roles in astrophysics. The X-ray background (XRB) constitutes another section of quasi-isotropic radiation flux reaching the Earth. It was discovered also in the 60's, and in the beginning its origin was enigmatic. In the next 2 or 3 decades the nature of the XRB was revealed. In the present article I would like to describe the basic characteristics on the XRB and recall a few examples of its relevance to cosmology.

2. The Extragalactic X-ray Background

The spectral energy distribution of extragalactic electromagnetic background radiation is schematically shown in Fig. 1. The signal defined as νB_ν allows for visual comparison of relative amplitudes generated in different energy bands. Apart from a clear peak due to CMB, other features are rather inconspicuous. In particular, this observation applies to the X-ray domain, although some distinct variations of spectral slope within that band are present.

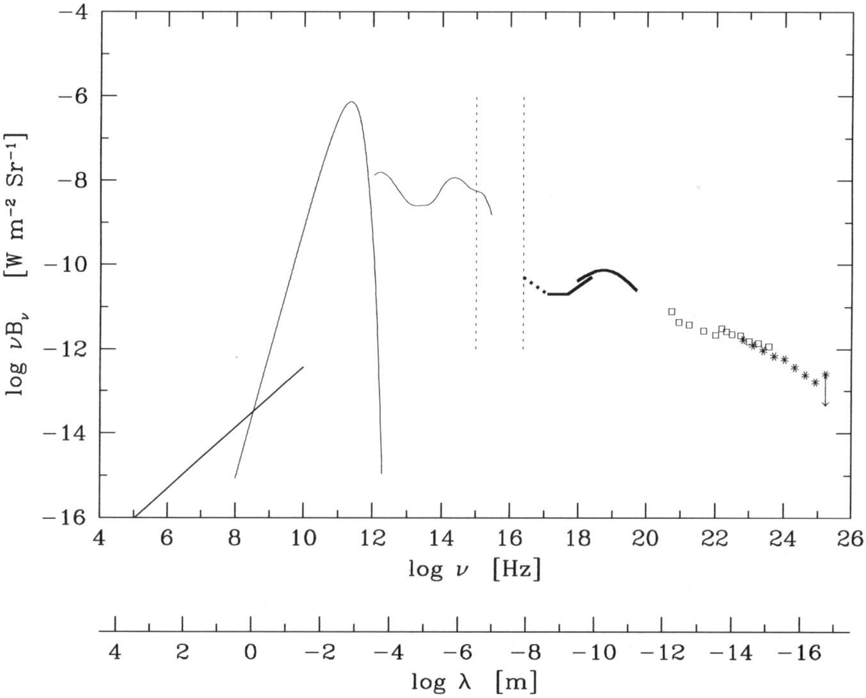

Figure 1

Spectral energy distribution of the extragalactic background. The data in the radio domain are based on the paper by Vernstrom et al. (2011); the CMB peaked between 10^{11} and 10^{12} Hz is represented by the black body with temperature of $T = 2{,}73$ K; in IR, optical and UV range the curve based on Mazin & Raue (2007) represents integrated galaxies emission; soft and hard X-ray data (thick curves) are from Miyaji et al. (1998) and Türler et al. (2010), respectively; data above 10^{20} Hz are from Inoue et al. (2009). Two vertical dotted lines indicate a spectral range inaccessible for extragalactic observations due to the strong absorption of UV and X-ray photons by neutral gas in the Milky Way.

It appears that only the CMB constitutes a truly diffuse component. In all the remaining domains, the signal is at least partially produced by discrete sources. In the X-ray band most of the signal generate active galactic nuclei (AGN). In soft X-rays (below ~ 5 keV) additional contribution of several (~ 10 percent) comes from hot gas in clusters of galaxies. Below ~ 1 keV some purly defined fraction of the XRB is produced by the so-called Warm-Hot Intergalactic Medium (WHIM).

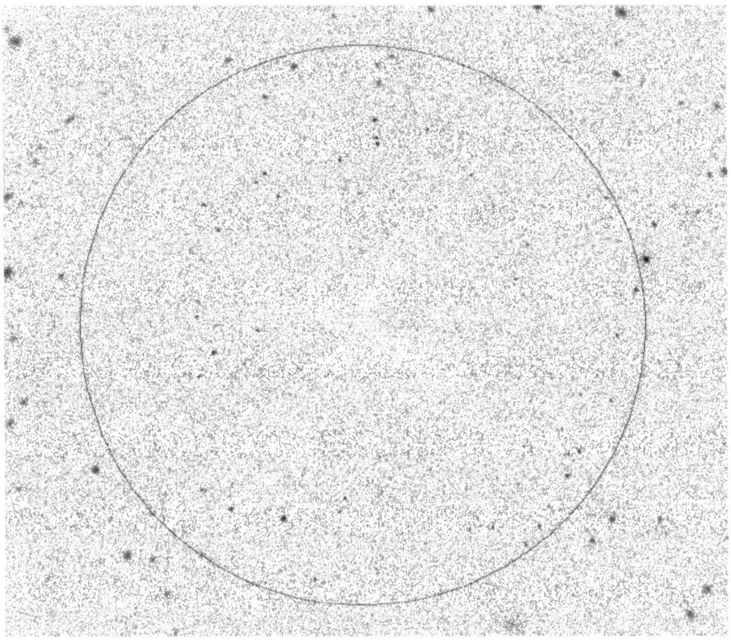

Figure 2

Fragment of *Chandra* Deep Field South in 0,5–2 keV band. Circle of 5 arcmin radius is shown. The apparent lack of strong sources in the central region is not real. It results from the strong degradation of Point Spread Function with the off-axis angle increase. Images of point-like sources far from the telescope optical axis are large, while sources close to the axis are contained within a couple of pixels in the figure and thus this makes them barely visible.

To resolve the X-ray background one needs a high angular resolution which is attained with imaging telescopes operating at energies typically below ~ 10 keV. At still lower energies, below 2 keV, the cosmic signal is relatively strong as compared to noise generated mostly by particle background. Deep exposures using the *Chandra*

X-ray Observatory[1] have resolved roughly 90% of the XRB in the energy band of 0,5–2 keV. In other energy bands the resolved fraction is smaller: below 1 keV – because substantial signal is in fact diffused, while at higher energies – because of noise, only relatively strong sources are recognized as individual entities. In Fig. 2 *Chandra* Deep Field South 2 Ms exposure in 0,5–2 keV energy band is shown.

In the present article I will discuss in the cosmological context the collective properties of three major components of the XRB, i.e. AGNs, clusters, and WHIM. Although each XRB fraction will be described separately, a common term 'background' was used in the title. This is partly because of historical reasons but also because procedures to evaluate statistical properties of the whole class of objects inevitably deal with the entire XRB, and – in some cases – methods applied in the investigation are characteristic for the continuous distribution rather than to the discrete one.

Active Galactic Nuclei

A central region of a galaxy exhibiting various kinds of peculiarities that cannot be attributed to a population of stars, is dubbed an Active Galactic Nucleus (AGN). In many instances, the nuclear activity is so prominent that it overshadows normal galaxy emission. The generally accepted model of the AGN assumes that in the galaxy center resides supermassive black hole (SMBH). As a result of gravitational interaction, the surrounding galactic material, gas clouds as well as individual stars, can be accreted by SMBH. During the accretion, substantial amount of gravitational energy is released. Although many details of the processes taking place in AGNs still remains obscure, it is believed that accretion onto the SMBH is the ultimate source of energy for all the species of the AGN zoo.

---

[1] See http://chandra.harvard.edu.

Two prominent classes of AGNs, viz. Seyfert galaxies and quasars, were identified as powerful X-ray sources in the 1970's, i.e. shortly after the construction of X-ray telescopes adequate for extragalactic research. Although both these types of galaxies were recognized earlier in the optical as bizarre objects, distinctly different from normal galaxies, discovery of their X-ray emission really added momentum to the investigation of all the AGNs, and helped to identify the 'central engine' common for the whole population. Extremely high bolometric luminosities (in fact, quasars emit most energy in the IR-optical-UV bands) imply a large accretion rate by a typical SMBH. The accreted matter accumulates in the black hole. Simple calculations (see below) show that the amount of mass integrated over the AGN lifetime is quite large (Sołtan 1982). Since the number of quasars in earlier epochs was substantially larger than now, a natural question arises: 'where have all the quasars gone?'

The reason why we do not see quasars in our proximity is that the typical accretion rates in the local Universe are much too small to generate prominent nuclear activity in the neighboring galaxies. The Milky Way is a good example of this situation. In the center of our galaxy resides the $4.4 \cdot 10^{6} M_{sun}$ black hole (e.g. Genzel et al., 2010) and it is barely visible in the X-ray band.

A question of local black hole demography is answered by explicitly using the assumption on the homogeneity of the Universe. Let L denotes bolometric luminosity of a single AGN. The accretion model postulates that luminosity is generated solely at the expense of the gravitational energy of the accreted matter. Thus

$$L = \frac{d\varepsilon}{dt} = \kappa \frac{d\mu}{dt} c^2,$$

where ε and μ are energy and mass of matter falling into a black hole, κ is the efficiency of mass to emitted energy conversion, and c is the speed of light. The relationship between the AGN instantaneous luminosity L and accretion rate $d\mu/dt$ integrated over time and summed over all SMBHs within a unit volume gives a relationship between the total energy E emitted by AGNs during the lifetime of the Universe and the mass M accumulated in SMBHs:

$$M = \frac{1}{\kappa} = \frac{E}{c^2}.$$

Now the assumption of the large-scale homogeneity of the Universe is applied: energy density E emitted by AGNs is independent of position (Cosmological Principle). Thus, it should be equal to the present local radiation energy density u, which in turn is related to the total flux f from AGNs received at Earth:

$$f = \int S n(S) dS$$

where S is the source flux, and $n(S)$ – the source number counts. However, the Universe expands and photon energies are reduced due to redshift z, what makes $u < E$. To correct for this effect one should introduce a $1+z$ factor (Sołtan 1982):

$$E = \frac{4\pi}{c} \int dS S n(S)(1+\bar{z}_s),$$

where $\bar{z}_s$ is the average redshifts of sources with the observed flux S. Substituting all the relevant quantities into the above formulae, one gets the total mass M that is distributed in a priori unknown number of SMBHs. However, assuming that just a small fraction of normal galaxies harbor the SMBH, one gets uncomfortable large masses for individual objects. To avoid the excessive mass problem, one needs to partition M into the majority (virtually all?) of galaxies. Consequently, the SMBH seem to be abundant in the local Universe.

Number source counts as a function of observed flux (in any spectral band) have demonstrated in the past to be one of the most effective cosmological tests. In the static, Euclidean, homogeneous, and non-evolving universe, the number of sources rises with decreasing flux S as $S^{-3/2}$. Any deviation from the '-3/2' power law indicates the departure of the real Universe from the assumed model. For instance, expansion of the Universe reduces the slope, while the cosmic evolution of sources often steepens the counts.

Strong cosmic evolution of AGNs has been discovered using radiosources identified with quasars and radiogalaxies (e.g. Schmidt 1968). In the late 1970's and 80's, analogous results were obtained by means of the extragalactic X-ray sources. The X-ray source counts, covering now more than 5 decades of flux, exhibit pronounced slope variations that reveal the cosmic evolution of AGNs as well as changes in the relative contribution by AGNs and normal galaxies.

Clusters of Galaxies

Clusters of galaxies are the most massive gravitationally bound structures in the Universe and from that reason alone are cosmologically interesting. However, for a long time the question of actual cluster masses has been shrouded in mystery. Zwicky (1937) proposed a systematic approach to the cluster mass problem. Soon after, it became clear that masses based on virial theorem are by a factor of 5–10 larger than masses determined using galaxy luminosities. The controversy was settled by means of X-ray observations. In the 70's it was found that rich clusters of galaxies are extended X-ray sources[2]. The emission mechanism was quickly identified as thermal Bremsstrahlung by hot gas filling the intracluster space. The X-ray data alone were sufficient to determine the cluster gravitational potential and – eventually – the total cluster mass.

Rich clusters of galaxies often have a regular shape: both distribution of galaxies and gas surface X-ray brightness are azimuthally symmetric. It is highly suggestive that the whole structure of such an object is relaxed and remains in hydrostatic equilibrium. Under these assumptions, a relationship between the gas pressure P_g and the mass is given by:

2 Clusters of galaxies were the first extragalactic objects identified with X-ray sources.

$$\frac{P_g}{dr} = -\frac{GM(r)}{r^2}\rho_g,$$

where $M(r)$ denotes total (gravitational) mass within radius r, ρ_g is the gas density and G is the gravitational constant. The gas equation of state is represented by perfect gas

$$P_g = \frac{\rho_g k T_g}{\mu m_H},$$

where T_g is gas temperature, k – Boltzmann constant, μ – mean molecular weight, and m_H – mass of hydrogen atom. These two equations, accompanied by the radial distribution of gas temperature, allow to find $M(r)$. For the isothermal gas, what is usually an adequate approximation, one gets:

$$M(r) = -\frac{kT_g}{G\mu m_H}\left(\frac{d\log\rho_g}{d\log r}\right)r.$$

Thus, the radial total mass function depends on the gas density profile that can be determined from the surface brightness distribution.

Results obtained from the X-ray measurements were generally in agreement with the virial mass determination. High cluster masses were later confirmed also by the weak lensing effect. It is now widely accepted that roughly 80 percent of the matter density of the Universe is non-baryonic and resides in the form of 'dark matter'. It is a collisionless gas that is detectable exclusively through gravitational interactions. As the chemical composition of dark matter is unknown, its nature remains a great mystery.

Missing Baryons

A major fraction of baryons in the present Universe escapes detection – it is 'invisible'. The visible matter in the local Universe resides in galaxies in the form of stars as well as gas and dust clouds. It may be detected in the optical or IR either in emission (stars, bright nebulae) or absorption (neutral gas and dust). A noticeable amount of baryonic matter is contained in hot gas in clusters of galaxies. The presence of this phase of matter is revealed by X-ray observations. All forms of 'luminous' baryonic matter in the local Universe sum up to roughly $\Omega_{\rm lum} = 0.015 - 0.02$, where $\Omega_{\rm lum} = \rho_{\rm lum}/\rho_{cr}$, and $\rho_{\rm lum}$ is the density of all matter detected via electromagnetic processes, and ρ_{cr} is the critical density. Present value of $\Omega_{\rm lum}$ is approximately a factor of 2 below the density parameter of all the baryonic matter $\Omega_{\rm b}$, estimated from primeval abundances and from Ly-α forest at redshifts $2-4$. The observation that $\Omega_{\rm lum} < \Omega_{\rm b}$ known as missing baryon problem, is not to be confused with the dark matter question, i.e. that baryonic matter itself constitutes a small fraction of the total matter density.

It appears that half of all the baryons still reside in the intergalactic space. Hydrodynamical simulations by Cen & Ostriker (1999) and other groups were used to investigate the evolution of the primordial gas occupying regions outside large concentrations of mass, i.e. galaxies and clusters. Rarefied plasma in the intergalactic space gradually accumulates in potential wells created by the existing concentration of matter, predominantly non-baryonic, or 'dark'. The motion of the gas towards galaxies depends on the distribution of the total gravitational potential and on the non-gravitational heating supported by so-called feedback, i.e. a transfer of energy from galaxies to the intergalactic medium. The simulations have shown that the infalling gas interacts with matter expelled from galaxies and shocks are generated. Shocks heat plasma around galaxies to temperatures $10^5 - 10^7$ K, while the material more distant remains at the temperatures of $\sim 10^4$ K (e.g. Davé et al., 2001). Since the heated plasma is to some extent compressed, it comprises a relatively well defined kind of intergalactic plasma, called Warm-Hot Intergalactic Medium (WHIM). It emits X-ray photons via free-free transitions. However, the plasma densities are still quite small and the emission intensity is extremely weak. Furthermore, it is restricted to low energies, below ~ 1 keV. Thus, the WHIM emission constitutes a minute XRB component that may be easily confused with the thermal emission by clouds in the Milky Way. However, the strong spatial correlation of the WHIM with the galaxy distribution gives a unique opportunity to isolate the WHIM signal from the dominating XRB amplitude.

In a series of papers (see Sołtan (2006), and references therein), I investigated structure of the soft XRB in the vicinity of galaxies using the cross-correlation technique. Assuming that the WHIM emission surrounds galaxies in the local Universe, one should expect enhancements of the XRB flux close to field galaxies:

$$\delta\rho\,(\theta) = s \cdot P(\theta) + s \cdot w_{gg}\,(\theta) \cdot n_g,$$

where:

$\delta\rho\,(\theta)$ – average excess of the X-ray surface brightness at angular distance θ from a galaxy

s – X-ray brightness of the individual WHIM halo around galaxy

$P(\theta)$ – normalized surface brightness of the WHIM halo

$w_{gg}\,(\theta)$ – autocorrelation function of galaxies

n_g – surface density of galaxies surrounded by the WHIM halos.

The first term on the right-hand side represents the X-ray signal generated by a single halo of the WHIM surrounding the galaxy, while the second term accounts for the non-random distribution of galaxies. As a result of galaxy clustering (described by the autocorrelation function w_{gg}), the average XRB surface brightness around each galaxy is raised due to the emission associated with the excess concentration of galaxies. Fig. 3 shows the XRB surface brightness in two energy bands (in instrumental counts per second per square arc second) as a function of angular distance θ from a galaxy averaged over several thousands of galaxies selected in more than 150 pointing observations obtained with the XMM-Newton[3] X-ray telescope.

[3] See http://xmm.esac.esa.int/.

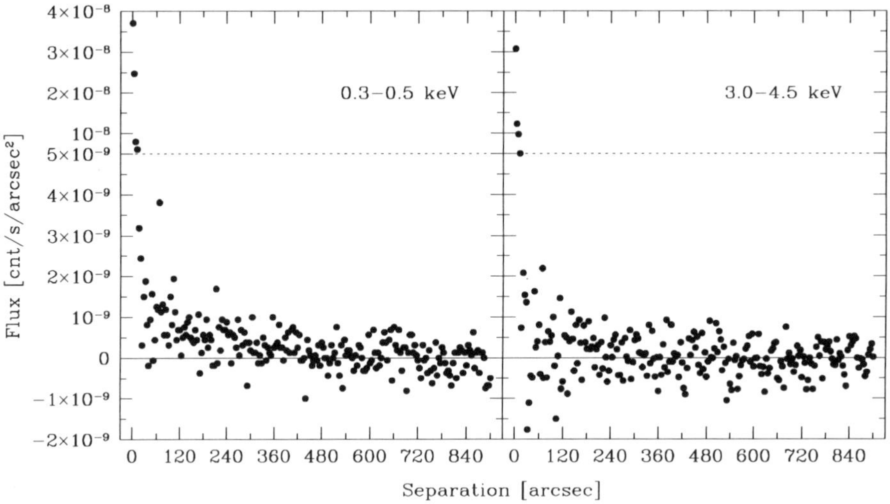

Figure 3

Cross correlation function of the galaxy distribution and the XRB in two energy bands: 0.5 – 5 keV and 3 – 4.5 keV. Plots give the average radial distribution of the XRB surface brightness around several thousands galaxies with magnitudes between 17 and 20 magnitude selected in ~ 150 X-ray fields. The X-ray observations were extracted from the XMM-Newton archives. To display the data points at the smallest separations, the scale at the *y*-axis above the dotted line has been compressed by a factor of 10.

In the soft (0.5 – 2 keV) and hard (3 – 4.5 keV) bands a strong signal is observed at the smallest separations (below ~ 15 arsec; note a change of ordinate scale). It reflects the fact that normal galaxies, not only AGNs, are X-ray sources. One should emphasize, however, that even this conspicuous signal is still weak as compared to the average amplitude of the XRB, which in both energy bands is close to 10^{-7} counts/s/arcsec^2. The emission inherent to the WHIM is expected to extend to larger separations In fact, it is easily visible in the soft band to 6 – 7 arcmin. The signal correlated with galaxies amounts here to 1 – 2 per cent of the total XRB flux, while in the hard it is barely visible to at most 3 – 4 arcmin and drops below the noise level at larger distances. It is essential for the WHIM question that the correlation between the galaxy sample and XRB is weak or absent at higher energies. It shows that the signal is not generated by galaxy clusters, because in that case it would be present in both energy bands.

Despite the fact that a huge data volume has been used in the analysis, statistical noise caused by limited numbers of galaxies and X-ray photons introduce large uncertainties into quantitative analysis of the WHIM. However, crude assessment of the parameters describing the part of 'dark baryons' is possible. The average size of the halo emitting soft X-rays exceeds 500 kpc and its mass amounts to $(3 - 7) \cdot 10^{11}$ M_{sun}. Clearly not all dark baryons are settled in the WHIM. Some baryons still form low density, relatively cold ($T \approx 10^4$ K) clouds populating the intergalactic space. Nevertheless, the X-ray observations reduce the fraction of baryons not accounted for by 8 – 9 percent.

3. Summary

The X-ray window was open for astrophysics 50 years ago. At that time cosmology was already a mature branch of science. Although numerous fundamental questions are still unanswered and new ones continue to arise, the X-ray observations have delivered a diversity of evidence which – in conjunction with other astrophysical data – helped to resolve some problems.

One of the revolutionary turns which changed our perception of galaxies as basic building blocks in the Universe, was realized as a result of X-ray observations of quasars and Seyfert galaxies. These two classes of extragalactic objects were known earlier, but neither optical, nor radio data provided any clue that SMBHs are responsible for virtually all kinds of nuclear activity in galaxies. The X-ray data have appeared to be crucial in the construction of a unified model of the AGN phenomenon. Dominating AGN contribution to the XRB helped to reevaluate spatial concentration of SMBH in the Universe. Since most galaxies play host to the SMBH, it is certain that galaxies as ensemble of stars and interstellar matter are inherently connected with SMBH, although the origins of this relationship remain unclear.

Clusters of galaxies were the first extragalactic objects clearly recognized as X-ray sources. The existence of strong discrete sources has strengthened the discrete explanation of the XRB. Originally, soon after the XRB discovery, a truly diffuse emission was considered. The detection of the X-ray emission from clusters has bolstered the idea of dark non-baryonic matter as a dominating component of the matter density of the Universe. Although virial theorem clearly visualized the problem, X-ray observations of hot intracluster gas allowed for quantitative investigation of the gravitational field within clusters and common acceptance of the reality of dark matter.

Until recently, the question of 'dark matter' has been accompanied by similar one of 'dark baryons'. This term was used to label a discrepancy between the density estimates of the baryonic matter in the early Universe and the present-day observations in the local Universe. X-ray observations have also alleviated the dark baryons problem. It was possible because in the soft X-rays, below 1 keV, a fraction of 'missing' baryons is not completely 'dark'. Hydrodynamical simulations indicated that a substantial amount of primordial matter still occupies areas not populated by galaxies. However, in the vicinity of galaxies, due to the interaction with galactic winds, the density and temperature of this matter are expected to increase. It appears that the X-ray observations confirmed these predictions. Weak halos of X-ray emission have been detected around field galaxies and it seems natural to ascribe this emission to the WHIM, i.e. some subset of baryons, ascertained earlier as 'dark'.

4. Acknowledgments

This work has been partially supported by the Polish NCN grant 2011/01/B/ST9/06023.

References

Cen R., Ostriker J.P., "ApJ" 1999, no. 514, p. 1.
Davé R., Cen R., Ostriker J.P., et al., "ApJ" 2001, no. 552, p. 473.
Genzel R., Eisenhauer F., Gillessen S., "RevMP" 2010, no. 82, p. 312.
Inoue Y., Totani T., Inoue S., et al., 2009, arXiv:1001.0103 [astro-ph. HE].
Mazin D., Raue M., "A&A" 2007, 471, 439.
Miyaji T., Ishisaki Y. Ogasaka Y., et al., "A&A" 1998, no. 334, p. L 13.
Schmidt M., "ApJ" 1968, 151, 393.
Sołtan A., "MNRAS" 1982, no. 200, p. 115.
Sołtan A.M., "A&A" 2006, no. 460, p. 59.
Türler M., Chernyakova M., Courboisier T.J.-L., et al., "A&A" 2010, no. 512, p. A 49.
Vernstrom T., Scott D., Wall J.V., "MNRAS" 2011, no. 415, p. 3641.
Zwicky F., "ApJ" 1937, no. 86, p. 217.

Andrzej Sitarz[1]
Jagiellonian University, Kraków
Copernicus Center for Interdisciplinary Studies, Kraków

Causality and Noncommutativity

1. Introduction

Within the last century our understanding of the world improved enormously both at the micro scale (particle physics) as well as at the macro scale (cosmology). However, there is still no theory which could offer a comprehensive view of all fundamental physics. On one hand we have gauge theories, with their roots in the mathematical theory of principal fibre bundles and connections, which nicely allows the formulation of classical and perturbative quantum field theory. On the other hand we have again a purely geometrical theory of gravity, mathematically set in the Riemannian (or pseudo-Riemannian) geometry, which gives no hints whatsoever as to its quantum version, the *mythical* quantum gravity. Both theories are well-tested and define the basics of our current knowledge. Yet, there is no unique and satisfactory theory (nor even an idea) how to combine both of them into one single mathematical framework.

So far, the attempts have been directed towards the modifications of General Relativity, so that it is presented like a gauge theory, which could lead to some quantum formalism. However, an attempt to look in the opposite direction, which originated some time ago from the work of Alain Connes, appears to be at least worth looking at. How does it

---

[1] Supported by the grant from The John Templeton Foundation

work? Instead of looking at gravity as some sort of gauge theory we should rather look at gauge theories as some modifications of pure gravity. However, in order to do so we need to enlarge our understanding of geometry using a set of new tools and methods, which goes under the name of *Noncommutative Geometry* (see [5, 11] for details).

2. The Language of Noncommutative Geometry

One of the main difficulties of noncommutative geometry is its language. In all modern physical theories, the fundamental concepts are always connected with the notion of *points* and *spaces*. This allows one to not only easily draw pictures and visualize many concepts, but also to include a very intuitive concept of locality. Even in the case of Quantum Mechanics, or Quantum Field Theory, when working with operators on Hilbert spaces, one very often can visualize them as local, or differential operators on $L^2(X)$, for a suitable space X.

In contrast, Noncommutative Geometry starts from a completely different point, namely an algebra and its representation(s) on a Hilbert space. Of course, just like a space alone is not sufficient to do geometry, we need also structure to be imposed on the algebra to work with some really interesting objects. The one, relevant for topology is that of C^* algebras:

Definition 2.1 Let $\mathcal{A}$ be an involutive Banach algebra (that is a complex[2] normed algebra, which is complete as a topological space in the norm). If:

$$||aa^*|| = ||a||^2,$$

then $\mathcal{A}$ is a C^*-algebra.

[2] All algebras we consider are over $\mathbb{C}$.

Example 1. If X is a (locally) compact Hausdorff space and $C(X)$ is the algebra of continuous functions on X, then $C(X)$ is a commutative (non) unital C^* -algebra. Such algebra is always abelian.

What is, however, more important, is the converse of the above statement, which forms the renowned Gelfand-Naimark theorem:

Theorem 2.1 (Gelfand-Naimark) *If a C^* algebra is commutative then the Gelfand transform, $\mu : \mathcal{A} \to C_0(X_{\mathcal{A}})$, where $X_{\mathcal{A}}$ is the Gelfand spectrum of $\mathcal{A}$ (space of homomorphisms from $\mathcal{A}$ to $\mathbb{C}$):*

$$\mu(a)(\chi) = \chi(a), \quad a \in \mathcal{A}, \chi \in X_{\mathcal{A}},$$

is an isometric $*$-isomorphism.

Any such map $\xi \in X_{\mathcal{A}}$ is called a character of the algebra, and, roughly speaking, characters of abelian algebras could be identified with points of the underlying space. Here begins the problem with the intuition, since this cannot be translated to the case of nonabelian algebras. These objects may have very few or even no characters at all, so we cannot think of them as functions on some space.

A bit of help (though certainly more abstract) comes from the second important fact:

Theorem 2.2 (Gelfand-Naimark-Segal) *Every abstract C^* -algebra $\mathcal{A}$ is isometrically $*$-isomorphic to a concrete C^* algebra of bounded operators on a Hilbert space $\mathcal{H}$. If the algebra $\mathcal{A}$ is separable then we can take $\mathcal{H}$ to be separable.*

So, instead of thinking of some abstract algebra we are able to make the notion very precise (though not in a unique way), identifying the algebra with an algebra (norm closed) of some bounded operators on a Hilbert space. A good example of C^* algebras is that of finite-dimensional ones:

Example 2 Any finite dimensional C^* algebras is of the form

$$M_{n_1}(\mathbb{C}) \oplus M_{n_2}(\mathbb{C}) \oplus \cdots \oplus M_{n_k}(\mathbb{C}),$$

for some $0 < n_1 \leq n_2 \leq \cdots \leq n_k$, $n_i \in \mathbb{Z}$.

Extending the Hilbert space to an infinite dimensional one (but with a countable basis) we arrive at the next-largest C^*-algebra (metaphorically speaking), which are compact operators. The algebra $\mathcal{K}(\mathcal{H})$ is a closure of the algebra of operators of finite rank (a closure of all finite-dimensional C^*-algebras) and it is a C^* algebra. Taking again a separable Hilbert space $\mathcal{H}$ it is not difficult to see that all bounded operators, $B(\mathcal{H})$, form a C^* algebra, which is the largest possible (by GNS construction) C^*-algebra.

To summarize this brief review we can present a dictionary showing corresponding terms:

SPACES	ALGEBRAS
(locally compact) topological space	C^*-algebra
homeomorphism	automorphism
continuous proper map	morphism
compact space	unital C^*-algebra
open (dense) subset	(essential) ideal
compactification	unitization
Stone-Čech compactification	multiplier algebra
cartesian product	tensor product

Noncommutative Geometry aims to extend this dictionary and equivalence to geometry, or more precisely to differential geometry and Riemannian geometry.

3. Connes' Way to Riemannian Geometry

Having a space, or even a topological space, or a manifold, is not enough for the purposes of physics. What one needs is a concept of a Riemannian manifold, that is a space equipped with a metric, which has suitable properties. Since we have no points there is no way to introduce the metric in the manner of prescribing the distances. Furthermore, since there is no generic construction of a tangent bundle or even the differential calculus of the first order, we cannot introduce (in a satisfactory way, at least) the metric tensor.

The way, which bypasses both problems has been to choose a construction which appears to be natural in the language we adapt and which contains in the classical situation all the necessary ingredients of Riemannian geometry. The construction is that of the Dirac operator, which might worry some people somewhat as it would mean that we need to restrict the notion only to the counterparts of spin or spin^c – geometries. Setting this apart (and keeping in mind that here are ways to circumvent the problem) we shall briefly recall here the main ingredients of the definition.

Definition 3.1 We say that a triple $(A,\mathcal{H},D)$ is a (unital) spectral triple if:

- $\mathcal{H}$ is a separable Hilbert space,
- A is a unital pre-C^*-algebra, faithfully represented as bounded operators on H, $\pi(a) \in B(\mathcal{H})$ for any $a \in A$,
- D is an unbounded and selfadjoint operator on H, so that its resolvent is compact on $\mathcal{H}$,
- For all $a \in A$ the operators $[D, \pi(a)]$ are bounded.

These are the most important conditions, which give us a *bare* spectral triple. Then we need to add some more conditions:

- Spectral triples are $\mathbb{Z}_2$-graded. We say that it is *even* if there exists an operator $\gamma = \gamma^\dagger$, $\gamma^2 = 1$, $[\gamma, \pi(a)] = 0$ for all $a \in A$ and $\gamma D + D\gamma = 0$. Otherwise we say that the spectral triple is *odd*.

• We say that the spectral triple satisfies *smoothness condition* if for all $a \in A$ both $\pi(a)$ and $[D, \pi(a)]$ belong to the domain of δ, $\delta(x) = [|D|, x]$.

• We say that the spectral triple satisfies a *finiteness condition* if the space of smooth vectors (intersection of domains of δ^m for all $m > 0$ is a finitely generated projective module over A.

• If there exists a nonnegative integer n such that the sequence of first k the eigenvalues of $|D|^{-1}$ (we ignore the kernel of D, which is finite-dimensional) arranged in a decreasing order and counting the multiplicities behaves like $k^{\frac{1}{n}}$.

• We say that a spectral triple is **real** if there exists an antilinear unitary J such that:

$$[J\pi(a)J^{-1}, \pi(b)] = 0, \quad \forall a, b \in A,$$

and

$$\left[J\pi(a)J^{-1}, [D, \pi(b)]\right] = 0, \quad \forall a, b \in A.$$

The last requirement is called the *order-one condition*. We say that the real spectral triple is of *KR*-dimension n if: the following conditions are satisfied:

$$DJ = \epsilon JD, \quad J^2 = \epsilon', \quad J\gamma = \epsilon''\gamma J,$$

where $\epsilon, \epsilon', \epsilon''$ are ± 1 depending on n modulo 8 according to the following rules:

n mod 8	0	1	2	3	4	5	6	7
ϵ	+	−	+	+	+	−	+	+
ϵ'	+	+	−	−	−	−	+	+
ϵ''	+		−		+		−	

• Finally, we say that the spectral triple is of *homological* dimension n, if there exists an n-Hochschild cycle with coefficients in the bimodule $A \otimes A^{op}$,

$$a_0 \otimes b_0 \otimes a_1 \otimes \cdots \otimes a_n = c \in Z_n(A, A \otimes A^{op}),$$

for which

$$\pi(c) = \pi(a_0)\left(J\pi(b_0)J^{-1}\right)[D,\pi(a_1)]\cdots[D,\pi(a_n)] = \begin{cases} 1 & \text{odd} \\ \gamma & \text{even} \end{cases}$$

There are numerous examples of spectral triples, starting with the ones presented originally with the definition [6], the examples coming from the so-called isospectral deformations [7]. A general theory of equivariant spectral triples [22] led to a family of real spectral triples on the noncommutative torus, which correspond to the different spin structures [20] and the construction of spectral triples on the Podles sphere [9] and the $SU_q(2)$ algebra [8]. The examples of locally compact geometries, like Moyal deformation, has also been incorporated as those admitting a spectral triple geometry. It should be mentioned, however, that not all of the above axioms can be met in all the examples and it is still uncertain which of them are essentially important.

4. Can we Have Lorentzian Noncommutative Geometry?

In principle, it is not very difficult to rewrite some part of the axioms for spectral triple for the case of an a arbitrary signature. The crucial problem, however, lies in the properties of the Dirac operator, which, unless like in an Euclidean case, fails to be an operator with a compact resolvent. Therefore one needs an additional ingredient of the theory, which corresponds to the choice of time orientation (in the easiest case of one time direction only).

The axioms, proposed in [21] are, in fact, analogous to those for the Euclidean case and are based on the idea that to each Lorentzian spectral geometry, it should be possible to associate a corresponding Euclidean one, as motivated by the classical situation [1].

Definition 4.1 *A geometric real (odd or even) Lorentzian spectral triple* of signature (1, q) is given by the data$(A, \pi, \mathcal{H}, D, J, \gamma, \beta)$, where:

• A is an involutive algebra, π its faithful, bounded, star representation on a Hilbert space $\mathcal{H}$,

• in the even case, $1+q \in 2\mathbb{Z}$, $\gamma = \gamma^\dagger$, $\gamma^2 = 1$ is a $\mathbb{Z}_2$ grading, commuting with the representation of A,

• J is an antilinear isometry such that:

$$[J\pi(a)J^{-1}, \pi(b)] = 0, \;\; \forall a, b \in A,$$

• $\beta = -\beta^\dagger, \beta^2 = -1$ is the $\mathbb{Z}_2$-grading (associated with the Krein-space structure), commuting with the representation of the algebra A,

• D an unbounded, densely defined operator, which is β-self-adjoint, that is:

$$D^\dagger = \beta D \beta,$$

and such that $[D, \pi(a)]$ is bounded for every $a \in A$, and $D\gamma = -\gamma D$.

• The operator

$$\langle D \rangle = \sqrt{\frac{1}{2}(DD^\dagger + D^\dagger D)} \tag{4.1}$$

has compact resolvent. In addition it is required that $[\langle D \rangle, [D, \pi(a)]]$ is bounded for all $a \in \mathcal{A}$.

• The grading, reality structure and the Dirac operator satisfy:

$$DJ = \epsilon JD, \quad J^2 = \epsilon', \quad J\gamma = \epsilon'' \gamma J. \tag{4.2}$$

where $\epsilon, \epsilon', \epsilon''$ are ± 1 depending on $1-q$ modulo 8 according to the following rules:

$1-q$ mod 8	0	1	2	3	4	5	6	7
ϵ	+	+	+	–	+	+	+	–
ϵ'	+	+	+	–	–	–	–	+
ϵ''	+		–		+		–	

• the Krein-space structure satisfies:

$$\beta\gamma = -\gamma\beta, \quad \beta J = -\epsilon^q J\beta,$$

• The Dirac operator satisfies the order-one condition:

$$\left[J\pi(a)J^{-1},[D,\pi(b)]\right]=0,\quad \forall a,b\in A.$$

• There exists a Hochschild cycle of dimension $n=1+q$, valued in $A^{op}\otimes A$,

$$c=b_o^{op}\otimes a_{i_0}\otimes a_{i_1}\otimes\cdots\otimes a_{i_n},$$

such that:

$$J\pi(b_o)J^{-1}\pi(a_{i_0})[D,\pi(a_{i_1})]\ldots[D,\pi(a_{i_n})]=\begin{cases}\gamma & n \text{ even}\\ 1 & n \text{ odd}\end{cases},$$

• We say that the Lorentzian spectral triple has time-orientation if there exist $a_i^o, a_i, b_i \in A$ such that

$$\beta=\sum_i J\pi(b_o)J^{-1}\pi(a_i)[D,\pi(b_i)].$$

If we do not assume the existence of J, we have a spectral triple without real structure but in such a case it is difficult to say whether we are in a typically Lorentzian case, with one *time-like* noncommutative direction, as we might as well be facing a geometry with a signature (p,q) where $p-q=1 \bmod 2$.

The above definition does not touch analytic properties (like summability, finiteness conditions etc), as these could be worked out using the operator $\langle D\rangle$ and will be similar to the ones of the Euclidean case (see [11], for instance).

Remark 1 The definition can be in a straightforward way extended to the case of arbitrary signature (p,q). Our notation is that the Euclidean spectral geometry of dimension 0 is identified with the $(0,n)$ signature.

The sign relations for arbitrary signature have been studied before in the context of *spectral geometry axioms* by various people [12, 16, 14].

5. Quantum Field Theory and the Concept of Causality

The classical concept of causality, that is the relationship between *cause* and *effect*, is based again on the notion of *points* which label the events taking place in space-time. However, using the language of algebras it might be relevant to start with the notion of *Einstein causality*, which has been developed for the purposes of Relativistic Quantum Field Theory. First, if we use the nets of algebras, that is algebras of quantum field theory operators, which are assigned to a region of space-time, we come across the following definitions of causality.

Let us also mention two related notions of macro- and micro-causality:

• **Einstein causality**: causal locality in QFT in terms of space-time indexed operator algebras $\mathcal{A}(O)$:

$$[A, B] = 0, A \in \mathcal{A}(O), B \in \mathcal{A}(O'),$$

for spacelike sparated regions

$$\mathcal{A}(O) = \mathcal{A}(O'')$$

causal shadow property; O'' is causal completion of O.

• **Macrocausality**: relates causality in QFT to some analytical properties of the S-matrix (see [15]), which classically reflects some properties of the retarded Green functions.

• **Microcausality**: is a simple condition for quantum fields evaluated at distinct points x and x', it is obeyed if:

$$[\Phi(x), \Phi(x')] = 0, \quad (x - x')^2 < 0.$$

for a quantum field Φ (understood as operators on a Hilbert space).

As we can see – even in the quantum field theory approach the basic underlying structure is that of a space-time, points and their space-like or time-like separation. For this reason, any field theory based on noncommutative geometry cannot be tested using any of the proposed properties of causality.

6. Challenges for Causality from Noncommutativity

Since there is no good notion of **points** and **locality** in noncommutative geometry and, moreover, the field theories on noncommutative spaces could be thought of as *nonlocal* ones, it appears that there is a long way before one could formulate a consistent notion of locality and – in the case of Lorentzian geometries – to formulate some principle of causality.

In most cases, the approach is similar as for the **nonclocal** theories, with the deformation parameter considered **small** so that there exists a **classical** limit. In the end, possible limitations of the theories are considered only in the perturbative expansions. This often leads to some contradictory results. Some consequences of the Lorentzian models have been discussed for the Moyal deformation, which is one of the best known noncommutative spaces, with the algebra generated by selfadjoint elements subject to the commutation relations of the quantum-mechanics type:

$$[x^\mu, x^\nu] = \theta^{\mu\nu}.$$

A.P. Balachandran et al. [2] observe that spacetime noncommutativity deforms statistics and therefore: 'generically violates causality in noncommutative quantum theories'. On the other hand, such violations are potentially a rich source of modifications of existing physics at the subatomic scale. Some potential features such as modification of the Pauli principle causing forbidden atomic transitions or correlations of observables in spacelike regions are mentioned (though without any specific analysis). Potential cosmological consequences also include anisotropies of the cosmic microwave background radiation.

A completely different conclusion has been reached by D. Bahns, in her work [3] she finds some remnants of causality at large distances: the cluster decomposition property, which remains intact on the noncommutative Minkowski space. Both approaches, however, use the interpretation of the underlying Euclidean space on which the noncommutative product is defined, as the space of *points* we observe, thus, in fact, treating the theory as a type of nonlocal quantum field theory. A treatment of the problem, which is based on the macro causality approach, come from the computations of perturbative quantum field theory for a scalar field on a Moyal space [4]. The results suggest, that, depending on the procedure assumed (as to the choice of which there are no independent indications) one has to give up give up causality or unitarity in noncommutative field theories.

As a sort of remedy, a different approach has been proposed by M.A. Soloviev [23], who proposed a concept of θ-locality, which could replace the micro locality condition listed above. Though this appears to be a relevant concept for the isospectral and Moyal-type deformations, it is limited only to those types of non commutativity.

Finally, let us mention that an alternative to *locality*, which is a so-called *wedge-locality*, has been proposed and studied for some of noncommutative models in [13]. The model contains a family of fields, labelled by different noncommutativity parameters, so that fields are related to each other by Lorentz transformations, which act also on the space of non commutativity parameters. Then one can create models, which obey the principles of covariance and locality, that is the fields localized in space-like separated wedges commute with each other.

7. A Proposal: Causal Structure on the States

The discussions of physical models concentrate around the Moyal deformation for an easy reason: this deformation allows presentation as a star product and in its properties is very similar to the usual, commutative geometry. In particular, the derivations, differential calculi, integrals are just the standard ones, moreover, the dual space (momenta) are completely commutative. Therefore there is no need even to depart from the usual scheme of thoughts – and there is no need for the approach using spectral triples.

However, if we consider some more general deformations, we need to start with the general setup, and so far, the one based on spectral triples appears to be the best suited for this purpose. Let us begin with the discussion of the metric, understood as the metric on the space of states $\mathcal{X}$: linear functionals from the algebra to $\mathbb{C}$. In the Euclidean spectral triple geometry one sets the distance between two states ϕ, ψ as:

$$d(\phi,\psi) = sup_{||[D,a]||\leq 1}|\phi(a) - \psi(a)|.$$

This leads, in the classical situation, to where one consider pure states, which correspond to points, to the usual metric on the space, which by the Gelfand-Naimark theorem corresponds to the commutative algebra. By definition such a metric is positive definite and as such cannot be used in the Lorentzian case. So far attempts to construct a Lorentzian version of the metric formula as presented above were limited to the classical situation. The first attempt was in [19], later extended to a more elaborate one in [17], however, they were not using the Dirac but rather the classical Laplace-Beltrami-D'Alembert operator.

We would like to indicate here that using the notion of Lorentzian spectral triples, as presented, we can go much further, using all building blocks of Lorentzian geometry, in particular, the operator β, which determines the time orientation.

Using β and the noncommutative integral (which is still an accessible tool even in the Lorentzian case, provided we use the Euclidean volume form) we can define the *spatial* elements of the algebra $A_s \subset A$ in the following way (n denotes here the dimension of spacetime and the metric dimension of $\langle D\rangle$):

$$a \in A_s \Leftrightarrow \fint \beta[D,a]\langle D\rangle^{-n} = 0.$$

Since in the classical situation β is a form dual to a global time-oriented vector field, the above orthogonality condition means that the gradient of a (which is computed through the commutation with the

Dirac operator) is *spatial*. Similarly, one might define elements of the algebra, which are *time-like*, $A_t \subset A$:

$$a \in A_t \Leftrightarrow \int\!\!\!\!- [D,b][D,a]\langle D\rangle^{-n} = 0, \quad \forall b \in A_s.$$

Then one can define the *spatial* distance between the states ϕ, ψ,

$$d_s(\phi,\psi) = sup_{||[D,a]|| \leq 1, a \in A_s} |\phi(a) - \psi(a)|,$$

as well as the *time-like* distance:

$$d_t(\phi,\psi) = sup_{||[D,a]|| \leq 1, a \in A_t} |\phi(a) - \psi(a)|.$$

Having both, it is rather easy to summarize that we can consider two states to be casually linked, if their time-like distance is bigger or equal their spatial distance. So, in fact, using the structure of Lorentzian spectral triples one is able to determine the causal structure on the space of states.

Of course, one should be aware of the possible shortcomings of the suggested definition. First of all, the definition might be ill-posed in some degenerate cases, where the metric dimension is 0. Furthermore, the spaces of spatial or time-like elements might be too small to really distinguish between the states. Finally, we should be aware that all this is relative to the choice of β, which plays the fundamental role in the definition of spectral triple and in defining the spatial elements of the algebra. Having a slightly different β (which is by no means unique) might result in a different subset of spatial and time-like elements and therefore in a completely different results for spatial and time-like distances.

We shall not discuss here the details of the definitions as well as its limitations and applications, leaving it for some future work. Certainly, one would want to require, for instance, that the causal structure on the space of states remains independent of β. Whether it is indeed the case, we can only conjecture at the moment. We hope that the above presented (or a similarly constructed definition of the causal structure of states) might be the basis of a better understanding of the relations between non commutativity and causality in physical models.

8. Conclusions

The models of noncommutative geometry pose a big challenge in the considerations of locality and causality. First of all, the notion of Lorentz (and Poincaré) symmetry might be deformed and we possibly cannot use the classical symmetries but rather must replace it with the Hopf algebras. Can we formulate a version of causality, which would correspond to the deformed symmetries?

Furthermore, most of our intuitions and approaches, also in the quantum field theory, are based on the notion of time as a continuous variable. Therefore, we need some time ordering methods, for example. Now, what can we do if the time is supposed to be a noncommutative variable? Can one allow for a oncommutative time at all?

Of course, all these questions have their classical counterparts. We barely grasp the true meaning of locality and causality at small distances or, for instance, in a situation of energy scales comparable to the Planck mass. A solution to these problems would be a major step towards reaching the more abstract ones, which are related to noncommutative geometry.

9. Acknowledgments

The author would like to thank the John Templeton Foundation for its support in the form of a grant.

References

[1] Baum H., *Spin-Strukturen und Dirac-Operatoren über pseudo-Riemannschen Mannigfaltigkeiten*, Teubner-Text zur Mathematik, Band 41, Teubner-Verlag, Leipzig 1981.

[2] Balachandran A.P., Anosh J., Padmanabhan P., *Causality and statistics on the Groenewold-Moyal Plane*, "Found. Phys." 2010, no. 40: 692Ð702.

[3] Bahns D., *Local counterterms on the noncommutative Minkowski space*, "Progress in Mathematics Volume" 2007, no. 251, pp. 11–26.

[4] Bozkaya H., Fischer P., Grosse H., Pitschmann M., Putz V., Schweda M., Wulkenhaar R., *Space/time non-commutative field theories and causality*, "Eur. Phys. J." 2003, no. C 29, 133Ð141.

[5] Connes A., *Noncommutative Geometry*, Academic Press 1994.

[6] Connes A., *Noncommutative geometry and reality*, "J. Math. Phys." 1995, no. 36, p. 619.

[7] Connes A., Landi G., *Noncommutative manifolds: The instanton algebra and isospectral deformations*, "Comm. Math. Phys." 2001, no. 221, pp. 141–159.

[8] Dabrowski L., Landi G., Sitarz A., Suijlekom W. van, Várilly J.C., *The Dirac operator on* $SU_q(2)$, "Comm. Math. Phys." 2004, no. 259, pp. 729–759.

[9] Dabrowski L., Sitarz A., *Dirac operator on the sandard Podles quantum sphere*, "Banach Center Publications" 2003, no. 61, pp. 49–58.

[10] Gayral V., Gracia-Bondia J., Iochum B., Schücker T., Varilly J., *Moyal planes are spectral triples*, "Comm. Math. Phys." 2004, no. 246, pp. 569–623.

[11] Gracia-Bonda J.M., Várilly J.C., Figueroa H., *Elements of Noncommutative Geometry*, Birkhäuser, Boston 2001.

[12] Gracia-Bondia J.M., private communication.

[13] Grosse H., Lechner G., *Wedge-local quantum fields and noncommutative Minkowski space*, JHEP 2007, no. 11, p. 012.

[14] Holfter A., *Nichtkommutative Geometrie und Quantisierung von Raumzeiten und Konfigurationsräumen*, Ph. D. Thesis, Mainz, 2003.

[15] Iagolnitzer D., *Causality in local quantum field theory: some general results*, "Comm. Math. Phys." 1992, Volume 144, no. 2, pp. 235–255.

[16] Meyer R., *Real spectral triples and charge conjugation*, [in:] *Noncommutative Geometry and the Standard Model of Elementary Particle Physics*, ("Hesselberg Academy", 14–19 March 1999) eds. F. Scheck, H. Upmeier, W. Werner, Lecture notes in physics, vol. 596, Springer 2002.

[17] Moretti V., *Aspects of noncommutative Lorentzian geometry for globally hyperbolic spacetimes*, "Rev. Math. Phys." 2003, no. 15, pp. 1171–1217.

[18] Paschke M., *Über nichtkommutative Geometrien, ihre Symmetrien und etwas Hochenergiephysik*, Thesis, Mainz 2001.

[19] Parfionov G.N., Zapatrin R.R., *Connes duality in Lorentzian geometry*, "J. Math. Phys." 2000, no. 41, pp. 7122–7128.

[20] Paschke M., Sitarz A., *On spin structures and Dirac operators on the moncommutative torus*, "Lett. Math. Phys," 2006, no. 77, 3, pp. 317–327.

[21] Paschke M., Sitarz A., *Equivariant Lorentzian spectral triples*, ArXiv: math-ph/0611029.

[22] Sitarz A., *Equivariant spectral triples*, [in:] *Noncommutative Geometry and Quantum Groups*, eds. P.M. Hajac, W. Pusz, Banach Centre Publications 61, IMPAN, Warszawa 2003, pp. 231–263.

[23] Soloviev M.A., *Noncommutative deformations of quantum field theories, locality, and causality*, "Theoretical and Mathematical Physics" 2010, no. 163(3): 741Ð752.

[24] StrohmaierA., *On noncommutative and semi-riemannian geometry*, "J. Geom. Phys." 2006, no. 56, pp. 175–195.

Michael Heller
Copernicus Center for Interdisciplinary Studies, Kraków
The Pontifical University of John Paul II, Kraków

Bottom-Up Causality in a New Setting

1. Causality in Physics

There is a quite popular view that physics says nothing about causality; it only determines temporal sequences of events. Such a view is often accompanied by saying: 'As Hume has long ago demonstrated…'. A more detailed account could look like that: Our experience, be it scientific or every-day experience, can only inform us about the contiguity, temporal succession and a constant conjunction of what we call cause and effect, but the very causal nexus between them is something that goes beyond our experience and is but an imposition of our mental categories (or simply habits) on what we observe and can measure. This is the view of a naïve empiricist, which, however, remains rather faraway from the real practice of physics. Physics is an empirical science but, far from being naïve, it uses highly sophisticated methods that are based on a strange correspondence between some mathematical structures and the structure of the world. This correspondence is 'strange' because it could not be expected a *priori*, although it is strongly validated by unprecedented successes in explaining how the universe functions. Because of this correspondence, at least some inferences within mathematical structures correspond to some causal dependencies observed in the world. The inferences within a given mathematical structure are purely formal things, but they model the real dependencies (also causal dependencies) between observed phenomena.

The above methodological analysis could be nicely supplemented by the practice of doing physics. Let us consider a neutron decay reaction, known in nuclear physics as the beta minus reaction. This reaction has been predicted from the theory of weak interactions which says that the down quark, the constituent of a neutron, transforms into the up quark, emitting an intermediating boson W^- that in turn decays into an electron and antineutrino. The process has been empirically verified.

The common-sense interpretation of this process states that there are subnuclear particles (quarks, neutrinos, antineutrinos, etc) that behave in a certain way, and clever physicists have identified, by ingenious guesswork, a mathematical structure which *describes* the process with a good approximation. However, physicists treat this process in a radically different way (if not in words then by their scientific practice). Elementary particles are but 'places' within the mathematical structure and outside of this structure have no meaning at all. The physical 'nature' of elementary particles is entirely determined by the mathematical structure. The beta minus decay is not *described* by mathematics; it is *prescribed* by it. If we make an analogy with software, we could say that elementary particles execute the program that is contained in a mathematical structure. Philosophers say sometimes that mathematical structures 'are lacking causal powers', and regard this as an argument supporting the Humean view on causality and against the Platonic interpretation of mathematical objects. As we have seen, the physical interpretation is just the opposite. 'Causal powers' that matter seems to come entirely from mathematical structures. There is nothing like matter possessing or not possessing certain properties. Empirical results, informing us about causal sequences of events are inseparably interwoven with mathematical structures forming together with them what could be called the stuff of the universe. Matter not only lacks 'causal powers'; without mathematical structures it simply does not exist.

Keeping this in mind we could now ask how the 'causal dynamics' of the cosmic stuff operates to form the highly structuralized world around us.

2. Modular Hierarchy of Causal Influences

Our macroscopic world is extremely complex: highly structured geological strata, a superbly diverse biosphere, animal organisms subordinated to the goal of survival, and a human brain whose complexity surpasses that of the rest of the universe by several orders of magnitude. There are strong reasons to claim that this unutterable richness is an outcome of something that mediates between the maximally simplex fundamental level and the level of our every-day experience. The fundamental level deserves to be called maximally simple because what we call complexity appears only on higher levels and because the physical laws that operate on it are in their, so to speak, purest form.

George Ellis explains that processes leading to emergent levels of complexity are based on the strategy of 'modular hierarchical structures'. 'The basic principle is that when you have a complex task to perform, you break it up into subtasks that are each simpler than the overall project, requiring less data and less computing power, and assign these tasks to specific modules. Each module is again split up into submodules until you reach a base level where the necessary tasks are simple operations that can be carried out by simple mechanisms'[1].

The modular structure of this strategy is essential. At each level the modules interact with each other forming a complex interaction network. Although each of them performs different tasks, the whole

---

[1] G.F.R. Ellis, *Top-down causation and emergence: Some comments on mechanisms*, "Interface Focus" 2012, no. 2, pp. 126–140; all quotations in this section come from this paper.

results into a new, highly structured system. Modules interacting at each level 'are relatively strongly bound internally with higher frequency and higher energy internal dynamics', and interact with other modules 'through weaker bounds and lower frequency interaction dynamics'. Every level of this hierarchy possesses a relative autonomy with respect to other levels: 'Each of the different levels of the hierarchy function according to laws of behavior appropriate to that level, and are describable only in terms of language suited to that level'.

Each level contributes to the entire hierarchical structure, but 'the real work is done' on the most fundamental level with each level 'feeding its results into the next higher level components until the desired result emerges at the appropriate higher level'. This is how bottom-up causality operates.

George Ellis emphasizes that there also exist causal influences acting in the opposite direction. Especially when very complex systems (such as a living cell) come into play, there is information that 'cannot be derived in bottom-up way because it implicitly embodies information about environmental niches'. In this sense, we have a kind of top-down causation which, by imposing boundary or initial constrains, influences lower hierarchy levels, at least those with which a given level directly neighbors.

Thus our macroscopic environment is shaped by interactions of bottom-up and top-down causal influences. It is clear that it cannot be understood without penetrating as deep as possible into the hierarchy of underlying levels.

3. Causality at the Fundamental Level

According to the present paradigm in physics, the most fundamental level (or the most fundamental level of which we are aware) is the so-called Planck level. In spite of the fact that so far there is no satisfactory theory of this level, we have several working models of it. Although using different mathematical methods and leading to different results, they generate more or less the same list of problems that have to be solved or at least elucidated by the future fundamental theory[2]. Let us quote some of them, especially those with are related to the issue of causality.

One of the greatest problems arising in the search for quantum gravity theory is the problem of space and time (of time perhaps even more so than of space). Almost all attempts to quantize gravity indicate that the space-time manifold, in its present form as a background or arena for physical processes, will not survive the quantum gravity revolution, there is a growing feeling among those engaged in this research that the future looked for theory should be background independent, that is to say that the physics of the fundamental level would not require any background to develop. In particular, space-time should emerge out of this primordial era as the result of some averaging processes.

If so, another dogma of classical physics, namely that all physical objects are 'well localized' in space and time, should be abolished. In quantum mechanics many effects related to entanglement, decoherence and delayed time experiments can be interpreted as testifying to the nonlocal character of the fundamental level. In cosmology the so-called horizon problem seems to point in the same direction. The fact that some physical properties, such as the temperature of the microwave background radiation on causally disconneced parts of the universe, are synchronized with a great preciseness, could be quite naturally explained if in the beginning (i.e. below the Planck threshold) everything was nonlocal, and only later on, because of the overall expansion, horizons emerged (together with other kinds of localities).

If then the fundamental level is background-free and nonlocal, what about dynamics? It seems that what we need here is a generalized

[2] For a list of such problems see: M. Heller, L. Pysiak, W. Sasin, *Fundamental problems in the unification of physics*, "Foundations of Physics" 2001, no. 41, pp. 905–918.

dynamics and since dynamics is indispensable elements of any causal nexus, we also need a generalized causality.

All the above signaled problems have a strong bearing on our understanding of causality. It is clear that until they are solved, or at least sufficiently clarified, our understanding of causality must remain incomplete and limited to not much more than macroscopic physics.

4. Ontology for Planck Level

To see how the discovery of the 'final physics' could affect our understanding of causality, let us try to reconstruct the 'ontology' of the Planck level based on our present, provisionary models. Philosophers have proposed several ontologies which, according to their claims, present the innermost structural architecture of our world. Some of them say that the world's building blocks are substances that have various properties owing to which they enter into various relationships between themselves (this is the ontology of the Aristotelian type); some others regard the universe as a process, and consequently prefer to speak of the universe as in the state of becoming rather than in the state of existence (an ontology of the Whiteheadean type). There is also a modification of this view according to which there are not processes but rather events that are 'ontological units' of the world. These events are structured by a net of relations between them, and this allows us to interpret the network as giving rise to what we call substances or processes.

The above sketched ontologies admit many modifications and variations, and all of them are evidently based on our macroscopic experience. In particular, all of them assume that objects (substances or events) are 'well localized' (in the process ontology time plays essential role, and time is composed of 'localized' events). If we seriously treat indication that the fundamental level is background-free and nonlocal, then these ontologies are clearly inadequate for this level. To be more concrete in our speculations on the ontology of the fundamental level, let us base our analysis on the assumption that this level is correctly modelled by the co-called noncommutative geometry. There are several such models, but I shall not stick to any particular of them. Instead, I shall base my considerations on general properties of noncommutative geometry that are valid, more or less, in all approaches using this mathematical theory. My choice of models based on noncommutative geometry is motivated by the fact that they are the most radical as far as nonlocality and the absence of a background are concerned[3].

The concept that is dominant our (macroscopic) concept of space is that of the point and its neigbourhood. Owing to this property we can localize objects in space, and in this way objects acquire their individualities. In algebraic language, points are identified with all smooth functions on a given space (such functions form in algebra) that vanish at a given point (such functions form what is called maximal ideal of a given algebra). Noncommutative spaces are defined in terms of noncommutative algebras, and such algebras, in general, have no maximal ideals. Consequently, the concept of point (and its neighbourhood) loses its meaning. Noncommutative spaces are thus nonlocal entities. A well localized object in such a space contradicts the formal structure constituting the noncommutative space.

Is physics possible in such circumstances? It is possible, but it must be adapted to the new conceptual environment. There are several approaches to construct such a new physics. Perhaps the most intuitive one replaces (in a sense) the concept of point with that of state. This concept has a clear algebraic definition and is entirely nonlocal. Just as in everyday situations, a crowd can be described as being in the state of a turmoil, and this description does not refer to single people but rather to the entire crowd. The difference between this situation

---

[3] A fundamental monograph on noncommutative geometry (difficult to read) is: A. Connes, *Noncommutative Geometry*, Academic Press, New York, 1994; an easier introduction: G. Landi, *An Introduction to Noncommutative Spaces and Their Geometries*, Springer, Berlin-Heidelberg, 1997. A particular model showing conceptual possibilities of noncommutative geometry: M. Heller, L. Pysiak, W. Sasin, *Conceptual unification of gravity and quanta*, "International Journal of Physics" 2007, no. 46, pp. 2494–2512.

and that in noncommutative geometry is that in the latter there are no single entities; only a global behavior has any meaning.

It turns out that if certain conditions are satisfied, even dynamics can be defined in terms of states[4]. In the standard, macroscopic physics dynamics is associated with the motion 'from point to point' and is measured by time. But these are local concepts and they are absent in the noncommutative environment. It is a surprise that essential features of dynamics (the form of dynamic equations) are preserved even if space and time (understood locally) are not available. A good analogy is when the usual dynamics is given in terms of integral vector fields (which are a global entities) rather than in terms of local differential equations.

It goes without saying that the problem of dynamics is but an aspect of the problem of causality. If we agree that causality is more than a simple sequence of events then it is just the dynamics that gives a causal nexus between cause and effect. Accordingly, in the noncommutative setting we are confronted with a generalized, nonlocal sort of causality. In this case we should forget about 'the cause here and the effect there'; the concepts 'here' and 'there' are not available. They are totally engulfed by global dependencies between various aspects of the physical world.

Nonlocal phenomena known from quantum physics (entanglements, decoherence, time delayed experiments) can be regarded as traces or vestiges that survived the transition from the Planck era.

5. Bottom-up Causality Revised

The role of bottom-up causality in physics is paramount. It is not only a cornerstone of the reductionist method in physics, but also seems

---

[4] See: A. Connes, C. Rovelli, *Von Neumann algebra automorphisms and time-thermodynamics relation in generally covariant quantum theories*, "Classical and Quantum Gravity" 1994, no. 11, pp. 2899–2917.

to be the necessary condition for top-down causality, As emphasized by Ellis, 'higher level conditions influence what happens at the lower levels, even if lower levels do the work'[5]. If, therefore, we want to penetrate more and more fundamental causal interactions, we should go to lower and lower levels of physics. As we, in this process, approach the Planck level, local concepts become more and more fuzzy and the global landscape gradually takes over. Finally, we reach the fully noncommutative regime with all its peculiarities. At the 'bottom' we have a purely nonlocal causality in which our familiar local dependencies are only virtually present as 'limiting cases'. If we agree that the main characteristic of top-down causality is a holistic type of explanation, and the main characteristic of the bottom-up causality is the reductionist type of explanation, then what happens on the fundamental level looks more like top-down than bottom-up causality: global properties are not reduced to local properties, but *vice versa* local properties are derived from global ones. I would say, it is a top-down causation with the reversed top-down arrow (now pointing from 'down' to 'up').

6. Instead of Conclusions

First of all, a warning: the noncommutative approach to the Planck level is not a well-established theory, but rather a hypothetical method (rather than a hypothetical theory or model) to look for a fundamental theory of physics, and it should be treated as such. Accordingly, this part of our analysis that refers to the application of noncommutative geometry to physics has only the value of 'playing with concepts'. However, it seems to be important at least for two reasons. First, it prepares us for the upcoming conceptual revolution in the foundations of physics that will certainly be associated with the discovery of the 'final theory'. Second, it has an important philosophical significance

[5] See reference 1, p. 127.

since it clearly demonstrates the limited character of some concepts which are rooted in our mentality. One of them is the concept of causality. We must be ready not only to modify (to enrich) our understanding of causality in physics, but also to be open for its broadening in philosophy. Let us think, for instance, about the First Cause of the universe. How should we formulate the question about the causal relationship between God and the universe? Should we start with the traditional concepts of cause and effect, which are strongly immersed in our local way of thinking, and try to analogously extend them to God and the universe? Or should we think, from the very beginning, in a global way? But even in this case we must be ready to admit that also this way of thinking might not be adequate if we go beyond the world of physics?

This is only one example, but philosophy abounds in reasonings in which causality plays a crucial role. In all these cases we should be ready to revise our conclusions. Open-mindedness is a criterion of a sound philosophy.

Mariusz P. Dąbrowski
Cosmology Group, University of Szczecin
Copernicus Center for Interdisciplinary Studies, Kraków

Varying Physical Constant Cosmologies and the Anthropic Principles

1. Introduction

Physics as an experimental science has been developed by formulating some mathematical laws which then were verified by measurements of the appropriate quantities entering these laws including the constants which were supposed not to vary. However, one is able to ask the fundamental question of why the laws we have formulated are of the form they are, and also why are the physical constants we have measured of the values they are? For example, we may ask why Newton's law is of the form that the force is inversely proportional to the second power of the distance between the masses? Why not to the third power or perhaps to some other fractional power? Surely, it is not forbidden in any way as a mathematical law, but does not apply to what we can see experimentally in our spatial neighborhood. We may also ask why the strength of the interaction between electric charges and masses diminishes with distance, while it grows with distance for quarks which carry color charges? Putting that a different way, the latter behave as if they obeyed an elastic force. Could we imagine gravity scaling like that, i.e., growing with distance. The answer is simple: perhaps yes – but not in any kind of our universe!

Another problem is that even if some of the laws which are suitable for our universe have a similar mathematical structure, they still give different quantitative output due to the factors they contain – physical constants. For example, Newton's law and Coulomb's law have the same inversely proportional dependence on the distance, but the constants of proportionality: the gravitational constant G, and the electric constant $k = 1/4\pi\varepsilon_0$, where ε_0 is the permittivity of space; are vastly different. This leads to extremely important consequences for a common life – we do not have to take care of the gravitational interaction while dealing with integrity of surrounding us daily objects like pieces of furniture etc., or common phenomena like friction since it is electric force which dominates. The same is true in particle accelerators, where gravity is usually negligible. However, we may imagine a different world in which constants G and k are such that the gravitational and electric interactions are comparable. Then, one would have to take them into account in all these considerations.

In view of the above statements, we immediately come to a philosophical problem of potentiality, i.e., the existence of a set of options to 'create' the universe in such a way and not in another way. In fact, we can imagine the whole set of different universes (or perhaps pieces of our universe) – each of them having totally different sets of physical laws and physical constants. In the theological context we may ask, if there was any freedom of the choice of physical constants and laws initially, and why they have been chosen in a way they are now?

More generally, we may investigate the problem of how different our world would have been if the laws and constants had had different values from what they are now in our piece of the universe? The next (and a very natural step) would be to raise the question about the evolution of mankind. Namely, are there any regions of space in which physics (laws and values of constants) does not allow our type of life? Do they allow any other type of life? And, even a more challenging question – what is life in the context of our knowledge of the physical laws and constants?

The content of this paper is as follows. In section II, we will discuss the roots of varying constants cosmologies. In section III we will review the theoretical realization of the varying constant scenarios, in particle physics models. Then, in section III we will show some examples of the anthropic coincidences and the anthropic principles which lead theoreticians to what will be mentioned in section IV – the multiverse concept. Finally, we will give some remarks.

2. Roots of Varying Constant Cosmologies

Physical laws contain physical constants. There are many physical laws which are accompanied by many physical constants. Physicists always tend to reduce the number of both. All the unification theories are based on the idea that the number of physical laws and constants should be as small as possible. A couple of years ago, there was a vivid and open discussion among the three eminent physicists Michael Duff, Lev Okun, and Gabriele Veneziano (Duff et al. 2002) about which of the constants of physics are really fundamental. The three of these constants seem to play the special role, since we obtain various physical theories keeping them non-vanishing or allowing to reach zero or infinity. This was visualized nicely in the form of the Bronshtein-Zelmanov-Okun cube (Okun 1991) which was made of the three fundamental constants such as: the gravitational constant G, the velocity of light c, and the Planck constant h. The axis of the cube are $[(1/c), h, G]$, and the classification of the physical theories is according to them having these constants to vanish or being non-zero values (Barrow 2002). The Newton's Mechanics (NM) is the center of the coordinate system ($G = h = (1/c) = 0$) which is the limit of the Theory Of Everything (TOE) or superstring theory ($G \neq 0, h \neq 0, (1/c) \neq 0$) (Polchinski 1998). The $(h, (1/c))$ plane contains the points corresponding to Quantum Mechanics (QM) for ($G = (1/c) = 0, h \neq 0$), Special Relativity (SR) for ($G = h = 0, (1/c) \neq 0$), and Quantum Field Theory (QFT) for ($G = 0, h \neq 0, (1/c) \neq 0$). The $((1/c), G)$ plane contains the points corresponding to Newton's Gravity (NG) for ($G \neq 0, h = 0, (1/c) = 0$), Special Relativity, and General Relativity for ($h = 0, G \neq 0, (1/c) \neq 0$). Finally, the ($h$, G) plane apart from NG and QM points, contains also the Newtonian Quantum Gravity (NQG) for ($G \neq 0, h \neq 0, (1/c) = 0$). However, it was then disputed between Duff, Okun and Veneziano (2002) that other choices (such as c and the fundamental string length λ, in TOE) are also possible. Then, bearing in mind what was said above, one should admit that the role of fundamental constants is crucial in physics.

Historically, the problem of an appropriate choice and of a measurement of the constants was one of the biggest challenges for physics as an experimental science. This especially refers to the measurement of the coupling constants of the fundamental interactions in nature. Nowadays, we know that there are four such interactions which are supposed to be unified in TOE: gravitational, electromagnetic, nuclear strong, and nuclear weak. The first step towards measuring one of these fundamental constants was made by Sir Henry Cavendish already in 1798. He measured the gravitational constant $G = 6{,}67 \times 10^{-11}\ m^3/kg\ s^2$ in the so-called Cavendish experiment which contained two big metal balls pulling the small ones, and the effect was measured by a torsion balance. As for the other physical constants the important step was the measurement of the speed of light c. Ole Rømer in 1675 found that $c = 220{,}000\ km/s$ which we know was too small, then followed by James Bradley in 1826 who measured it as $c = 301{,}000\ km/s$, and Albert Michaelson in 1926 who found that $c = 299{,}796\ km/s$.

Nowadays, we use the International System of Units (SI) which contains seven units for the basic physical quantities (meter, kilogram, second, ampere, kelvin, mole, candela) out of which all the other units are supposed to be derived (Uzan 2003). However, already in the 19th century eminent physicists started to think of a basic set of physical units ('natural' units) of which all the other physical units could be derived. George Johnstone-Stoney in 1874 (Johnstone-Stoney 1881) introduced his 'natural' units of charge, mass, length, and time as:

the 'electrine' unit $e = 10^{-20}$ ampere-seconds (ε_0 – permittivity of space in the Coulomb's law),
the mass $M_J = (e^2/4\pi\varepsilon_0 G)^{1/2} = 10^{-7}$ gram,
the length $L_J = (Ge^2/4\pi\varepsilon_0 c^4)^{1/2} = 10^{-37}$ meters,
the time $t_J = (Ge^2/4\pi\varepsilon_0 c^6)^{1/2} = 3 \text{ x } 10^{-46}$ seconds.

This was later modified by Max Planck in 1899, following the discovery of the electron charge by J. Thompson in 1897 as $e = 1.6 \text{ x } 10^{-19}$ Coulombs, the introduction of the Planck constant $h = 6.6 \text{ x } 10^{-34}$ $J\,s$, and the Boltzmann constant $k = 1.38 \text{ x } 10^{-23}$ J/K into:

the Planck mass $M_{pl} = (hc/G)^{1/2} = 5.56 \text{ x } 10^{-5}$ gram,
the Planck length $L_{pl} = (Gh/c^3)^{1/2} = 4.13 \text{ x } 10^{-35}$ meters,
the Planck time $t_{pl} = (Gh/c^5)^{1/2} = 1.38 \text{ x } 10^{-43}$ seconds,
the Planck temperature $T_{pl} = (hc^5/k^2G)^{1/2} = 3.5 \text{ x } 10^{32}$ Kelvin.

At the beginning of the 20th century, the studies of ratios between various physical constants were initiated. Hermann Weyl (1919) found that the ratio of the electron radius to its gravitational radius was about 10^{40} (Weyl 1934). The topic was also studied by Arthur Eddington (Eddington 1923, 1934) who found the ratios: 1) proton-to-electron mass $1/\beta = m_p / m_e \sim 1840$; 2) an inverse of the fine structure constant $1/\alpha = (hc)/(2\pi e^2) \sim 137$; 3) ratio of electromagnetic to gravitational force between a proton and an electron $e/(4\pi\varepsilon_0 G m_e m_p) \sim 10^{40}$; which lead him to the introduction of the 'Eddington number' $N_{edd} \sim 10^{80}$. However, his explanations of the numbers found became too 'numerological' in the sense that he started to concentrate on some 'magical' explanations of them. For example such a number as $1/\beta$ was considered as ratio of the roots of an elementary equation like $10\,m^2 - 136\,m + 1 = 0$ or another ratio could be derived as $1/\alpha = (1/2)(16^2 - 16) + 16$ (Barrow and Tipler, 1986).

Eddington's ideas were put into a deeper physical context in Dirac's Large Numbers Hypothesis, who in February 1937 (while on his honeymoon with Eugene Wigner's sister Margit), made *inter-*

esting remarks about the relations between atomic and cosmological quantities (Dirac 1937). Firstly, as with Eddington, he studied the ratio of electromagnetic to gravitational force between a proton and an electron to be

$$\frac{F_e}{F_p} = \frac{e^2}{4\pi\varepsilon_0 G m_e m_p} = 2,3 \cdot 10^{39} \sim 10^{40}, \tag{1}$$

but instead of only concentrating on the 'mystery' of the number, he asked the physical question of why electromagnetic force is so strong compared to the gravitational? He then turned to cosmology and calculated the ratio of the observable universe to the classical radius of the electron (in order to get it one compares the electric potential energy of a ball of a constant density charge to its relativistic energy), and found that:

$$\frac{c / H_0}{e^2 / 4\pi\varepsilon_0 m_e c^2} = 5,1 \cdot 10^{40} \sim 10^{40}. \tag{2}$$

As we know, the Hubble parameter at present is

$$H_0 = 72\, km / s\, Mpc \; (1 \text{ parsec } (pc) = 3,0856 \cdot 10^{18}\, cm). \tag{3}$$

Dirac also calculated the number of protons in the observable universe to be

$$N = \frac{4}{3} \frac{\pi \rho (c / H_0)^3}{m_p} = \frac{c^3}{2 m_p G H_0} = 5,55 \cdot 10^{80} \sim N_{edd} \sim (10^{40})^2 \tag{4}$$

where the critical density $\rho = (3H_0^2)/(8\pi G) \sim 10^{-29}\ g/cm^3$. As it was already known from Einstein, Friedmann, Lemaître, and Hubble in the 1930's, the Hubble parameter was changing in time as $H(t) = [da(t)/dt]/a(t)$, where $a(t)$ was the scale factor describing the expansion rate of the universe. Then, from the relations (1) – (3), Dirac (1937) made an important remark: the ratio (2) is of the order of $\sqrt{N_{edd}}$ and the ratio (4) is of the order of N_{edd} and they contain the time-varying Hubble parameter, so these ratios must be either the result of a coincidence of the current moment of the evolution and they will change in time, or, if they are to stay the same because of some physical reasons, then the physical constants they contain (e, G, c, m_e, m_p) *would have to evolve in time.* Since the changes of e, m_e, m_p would need reformulation of the atomic and the nuclear physics, then Dirac chose a 'safe' option that only the gravitational 'constant' G would change in time according to the law

$$G \sim H(t) = \frac{da/dt}{a}\,, \tag{5}$$

so that $a(t) \sim t^{1/3}$, and $G(t) \sim \frac{1}{t}$. (6)

This led him to a very suggestive physical conclusion: electromagnetic force is so much stronger compared to gravitational force because the universe is very old, i.e., their ratio is proportional to the cosmic time (Dirac 1938)

$$\frac{F_e}{F_p} \sim \frac{e^2}{m_e m_p} t \sim t \tag{7}$$

which now is billions of years.

2. Theoretical Realization in Particle Physics Models

It was Ernst Mach who first suggested that the inertial mass of a given particle in the Universe is a consequence of its interaction with the rest of the (mass of the) Universe. Einstein hoped his theory adopted the requirement, but both the solutions of De Sitter (1917) – an empty universe evolving in time, and of Gödel (1949) – rotating universe, proved that it is not fulfilled in general relativity. However, the ideas of Mach were further incorporated in Jordan's conformal gravity (Jordan 1959) and especially in Brans-Dicke scalar-tensor gravity (Brans and Dicke, 1961), who assumed that the gravitational constant G should be associated with an average gravitational potential (scalar field) Φ surrounding a given particle, i.e., that there is an average potential

$$\langle\Phi\rangle = \frac{GM}{c / H_0} \sim \frac{1}{G} = 1{,}35 \cdot 10^{28}\ g / cm\,, \tag{8}$$

which is responsible for the gravitational force. In general relativity of Einstein – the gravitational constant is a constant (G = const.), and so the Einstein equations read as

$$R_{\mu\nu} - \frac{1}{2} g_{\mu\nu} R = \frac{8\pi G}{c^4} T_{\mu\nu}\,, \tag{9}$$

while in Brans-Dicke scalar-tensor (scalar field plus tensor field) gravity one should use the time-varying Φ field instead of G:

$$R_{\mu\nu} - \frac{1}{2} g_{\mu\nu} R = \frac{8\pi}{\Phi c^4} T_{\mu\nu} + \frac{\omega}{\Phi^2}\left[\Phi_{,\mu}\Phi_{,\nu} - \frac{1}{2} g_{\mu\nu}\partial_\rho\Phi\partial^\rho\Phi \right] + \\ + \frac{1}{\Phi}\left[\Phi_{;\mu\nu} - g_{\mu\nu}\Phi_{;\mu}^{\ ;\mu} \right] \tag{10}$$

where ω is the Brans-Dicke parameter. Einstein's general relativity theory is recovered for $\omega \to \infty$.

The idea is used in theories of unification of fundamental interactions such as the superstring theory, where the variability of the intensity of the interactions is a rule. There, we have the so-called 'running' coupling constants

$$\alpha_1 = \alpha \text{ (electromagnetic)}, \tag{11a}$$
$$\alpha_2 = \alpha_w \text{ (weak)}, \tag{11b}$$
$$\alpha_3 = \alpha_s \text{ (strong)}. \tag{11c}$$

As it is seen from Fig. 1, due to the high (or unbroken) symmetry in the Early Universe, only one common interaction existed and the values of the running coupling constants met at one point about 10^{16} *GeV* when the particles had high energies (Polchinski 1996). One of the most intensively studied model of particle physics in order to realize such kind of unification is the low-energy-effective superstring theory which is equivalent to the Brans-Dicke theory for $\omega = -1$. In such a theory, the so-called string coupling constant (which is 'running') $g_s = exp(\varphi/2)$, evolves in time with φ being called the dilaton, which is related to the Brans-Dicke scalar field by $\Phi = exp(-\varphi)$. The field equations are as follows

$$R_{\mu\nu} - \frac{1}{2} g_{\mu\nu} R = (\omega + 1)\partial_\mu \varphi \partial_\nu \varphi - \left(\frac{\omega}{2} + 1\right) g_{\mu\nu} \partial_\rho \varphi \partial^\rho \varphi + + g_{\mu\nu} \varphi_{;\mu}{}^{;\mu} - \varphi_{;\mu\nu} \tag{12}$$

and the variability of G is encoded in the variability of φ. What is interesting in such a simple superstring theory is the fact that a different choice of a beginning of the evolution of the universe than a big-bang is possible. This is called pre-big-bang cosmology of Gasperini and Veneziano (2003) in which the evolution begins at temporal minus infinity and a big-bang ($t = 0$) is a transient state (Fig. 2). In fact, there are four branches (options) of the evolution: two of them pre-big-bang (1 and 3), and two of them post-big-bang (2 and 4).

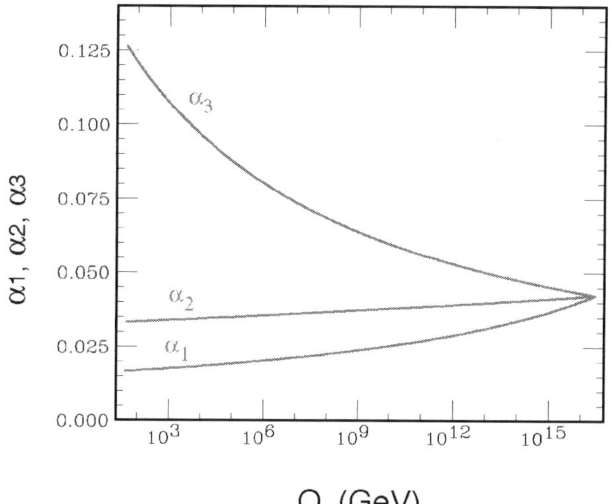

Figure 1

The unification of the coupling constants of the electromagnetic, weak and strong interactions at high energies.

Typical evolution contains branches 1 and 4 separated by the curvature and strong coupling (physical interaction $g_s \rightarrow \infty$) singularity at $t = 0$. This extension of the evolution onto the negative times encourages us to think of cyclic (or 're-birthing') universes which have their origin in ancient Greek tradition. One of the intriguing contemporary models are brane universes (Hořava and Witten 1996a, 1996b; Randall and Sundrum 1999a, 1999b) which lead to the idea of ekpyrotic and cyclic cosmologies (Khoury et al. 2001). For such models, a fine-tuned choice of an attractive (positive pressure) matter and a repulsive (negative pressure) matter results in infinite oscillations of the universe. In such brane universes the big-bang is repeated many times as a collision of two less-dimensional branes (thin films) in a higher-dimensional space.

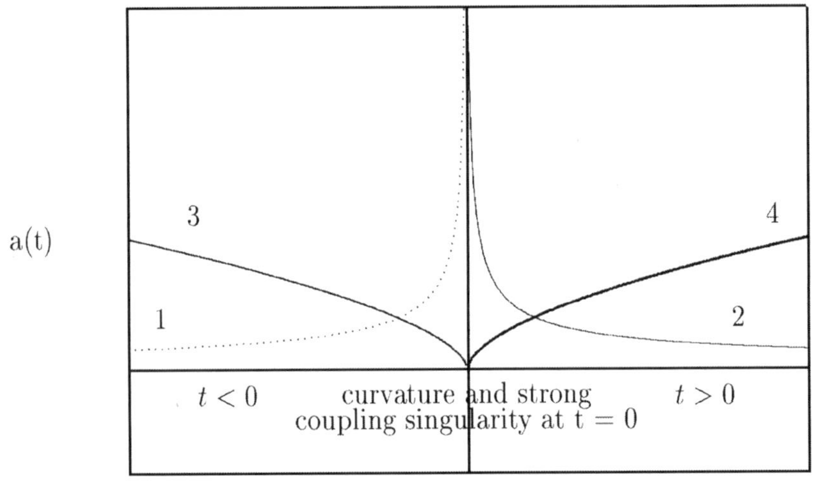

Fig. 2

Four possibilities for the evolution of the universe in pre-big-bang cosmological scenario based on the low-energy-effective action of superstring theory. It is usually chosen that the universe starts at minus infinity in time (branch 1) and evolves towards curvature and strong coupling singularity at t=0 (big-bang), finally evolving towards infinity (branch 4). In cyclic cosmological scenario a different pre-big-bang branch 3 is chosen in order to avoid a strong coupling singularity.

There are many other theoretical realizations of the varying physical constant theories. The most important of them are:

- varying speed of light theories (VSL): $c = c(t)$ (Albrecht et al. 1999, Barrow 1999, 2003);
- varying fine structure constant α theories: $\alpha = (2\pi e^2)/(hc) = 1/(137{,}036) \rightarrow \alpha = \alpha(t)$ (Magueijo et al. 2002, Mota and Barrow 2004, Barrow 2010);
- varying beta β (electron-to-proton mass ratio): $\beta = m_e / m_p = 1/1836 \rightarrow \beta = \beta(t)$
- varying weak/strong coupling constant theories: $\alpha_{w,s} = (g^2)/(2hc)$ (~ 1/27 , ~1) $\rightarrow \alpha_{w,s} = \alpha_{w,s}(t)$

There exist various observational constraints on the variability of the constants (for a review, see Uzan 2003). Some extremely selected examples are as follows:

1 Observational limit on $|(dG/dt)/G| < 9 \times 10^{-13}$/year – comes from primordial nucleosynthesis (Accetta et al. 1990);
2. Observational limit on $|(dG/dt)/G| < 1.6 \times 10^{-12}$/year – comes from helioseismology (Guenther et al. 1998);
3. Observational limit on $|(dG/dt)/G| < (4 \pm 9) \times 10^{-13}$/year – comes from lunar laser ranging (LLR) (Williams et al. 1996);
4. Observational limit on $\Delta\alpha/\alpha = (3.85 \pm 5.65) \times 10^{-8}$ – comes from Oklo phenomenon (Shlyakhter 1976, Petrov et al. 2006);
5. Observational limit on $\Delta\alpha/\alpha = (-8 \pm 16) \times 10^{-7}$ – comes from meteorite dating (long-lived beta decays) (Olive et al. 2003);
6. Observational limit on $\Delta\alpha/\alpha = (-0.5 \pm 1.3) \times 10^{-5}$ – comes from quasar absorption spectra with redshifts $2.33 < z < 3.08$ (Murphy et al. 2001);
7. Observational limit on $\Delta\beta/\beta = (5.7 \pm 3.8) \times 10^{-5}$ – comes from quasar absorption spectra (Ivanchik et al. 2005).

From these limits it emerges that physical 'constants' may vary, though very slightly. However, if taken into account throughout the whole evolution history of the universe, their variability may have some influence on physics now and then. On the other hand, if these constants varied stronger, then they could have had a vast influence on the view of our universe today, and perhaps made the life impossible.

3. Anthropic Coincidences and Anthropic Principles

We may ask a fundamental question: what is the place of man in the Universe? This can be understood both in a physical and a philosophical (or rather theological) sense. Let us take this in the physical context first. Observations of masses and sizes of the objects in our Universe says that they are not arbitrary – masses are proportional to sizes. Why are they not random? Why do only certain structures in the universe exist? Apparently, on the physical basis the answer is simple: they exist due to stable equilibria between fundamental interactions! For example common things like crockery or a book exist due to a balance between an attractive force between the protons and electrons and a repulsive force (pressure) of degenerated electrons.

Some other examples of coincidences for our life to exist are as follows:

- anomalous expansion of water – volume of water shrinks (density grows) at 0-4 Celsius degree which allows fish and other living creatures to survive the winter;
- no chaotic change of planets' obliquity (axial tilt) – the Earth has only one, and a relatively large natural satellite which is the reason that it does not 'wobble' chaotically. Mars has only tiny moons (captured asteroids ~ 10 km) and its chaotic change of obliquity can be as much as 45 degrees. The poles may almost swap places with the equator. The Earth without the Moon could change its obliquity of about 6-7 degrees while the Earth with the Moon changes it obliquity only of about 1,3 degrees – the Moon stabilizes the Earth (Laskar, Robutel 1993) which is vital for life to survive;

- way to a rich chemistry – electrons are light enough compared to the nuclei and so they allow the formation of complicated chemical structures by atoms. In fact, the size of the nucleous/size of an atom is about $\alpha\,\beta = (1/137)(1/1836) \ll 1$. If the size of a nucleus and an electron were comparable, then the atom would look like a binary system of stars, and the chemicals would be more difficult to form;
- primordial nucleosynthesis (formation of nuclei) is possible only for a very narrow range of values of α and β (see below);

and many others (see e.g. Barrow and Tipler 1986).

The last of the examples shows a very interesting relation between particle physics and cosmology, so we will say something more about it. Nucleosynthesis (formation of the nuclei made of protons and neutrons) in the early Universe ($0.04\,s < t < 500\,s$) may only take place in a fixed period of temperatures T (or times) which is driven by the values of the constants α and $\beta = m_e / m_p$, i. e.,

$$0.1\, m_e < T < \alpha\, m_p\,. \tag{13}$$

This interval of temperature could not have existed, if

$$\alpha < \beta\,. \tag{14}$$

Luckily, for an electron e and a nucleon

$$\alpha = 1/137 > \beta = 1/1836\,, \tag{15}$$

and so the nucleosynthesis is possible. However, if an electron had a mass equal to a muon μ, for example, then

$$\alpha = 1/137 < \beta = m_\mu / m_p = 0.11\,, \tag{16}$$

and the nucleosynthesis could not have been possible.

In fact, all the above statements are the examples of the anthropic principle that the Universe is extremely 'fine-tuned' to host us humans!

The Anthropic Principles

Let us first ask: what are the anthropic principles? Some say that they are tautologies – self-explanatory statements with no practical use.

They ask if the anthropic principles are 'physical' at all. Should physicists take them into account or just ignore them as trivial statements with no meaning for a contemporary scientific method of physics? We think that this is not so and will try to give some arguments that they should somehow be taken into account at least as potential options for the evolution of the physical universe. In fact, the anthropic principles were first presented in Kraków in 1974 by Brandon Carter during the IAU meeting (Carter 1974), and further developed by Barrow and Tipler (1986).

A fundamental question we face is as follows: what was the reason that out of many possible ways of evolution of the Universe a special one was chosen – the one which led to the formation of the galaxies, stars, planetary systems, and consecutively: an unconscious life and a conscious life? It is obvious to relate this question to many fields of science (physics, chemistry, biology, psychology) and to philosophy and theology. There are many specific formulations of these principles. We list the most elementary and the most characteristic ones below (in our discussion we will follow Barrow and Tipler (1986)).

Weak Anthropic Principle

It says that the observed values of all physical and cosmological quantities are *not equally probable*, but that they take on values restricted by the requirement that there exist sites where carbon-based life can evolve and by the requirement that the Universe be old enough for it to have already done so. In other words, life may have evolved – this is usually expressed by the application of the famous Bayes theorem in probability theory.

Strong Anthropic Principle

It says that the Universe *must have* the properties which allow life to develop within it at some stage of its history. In other words (or more physically) it says that the constants of Nature (e.g. gravitational constant) and laws of Nature (e.g. Newton's law of gravity) must be such that life can exist.

There are various interpretations of the strong anthropic principle and some of them are really curious. We will enumerate some of them below.

Interpretation A

It says that there exists only one possible Universe 'designed' with the goal of generating and sustaining 'observers'. It is no wonder that it is very teleological, and this is why it is sometimes also called 'An Intelligent Project' interpretation.

Interpretation B

It is very radical and says that observers are necessary to bring the Universe into being. Such a statement is based on the philosophical ideas of Berkeley. It was developed by John Archibald Wheeler in his PAP – Participatory Anthropic Principle.

Interpretation C

We think that it is the most attractive intellectually. It says that the whole ensemble of other and different universes is necessary for the existence of our Universe. The best-known, though most controversial version of this interpretation is the many-worlds theory of Everett (1957) in which everybody has many different branching histories of the evolution (DeWitt and Graham 1973). This was found interesting in the context of contemporary superstring theory (Polchinski 1998) which admits an incredible number of ways, i. e.,

10^{500} (one and five hundred zeros) – $(10^{100})^{5} = (1 \text{ googol})^{5}$

of generating possible sets of constants and laws of Nature (Douglas 2003, Susskind 2003) – this is called the superstring landscape. It is interesting that the number 1 googol = 10^{100} is a number first defined by Edward Kasner in 1938 and named by his 9 year-old nephew. Note that one googol is greater than the Eddington number N_{edd}. Curiously, there is even much greater number defined – 1 googolplex which is 10^{googol} (ten to the power of 1 googol) and contains the one, appended with one googol = 10^{100} of zeros.

Final anthropic principle

It says that intelligent information-processing must come into existence in the Universe, and, once it comes into existence, it will never die out.

Another formulation (of the final anthropic principle):

It says that no moral values of any sort can exist in a lifeless cosmology.

We will discuss one of the versions of the interpretation C for the strong anthropic principle in more detail. Suppose that W is a set of all possible universes with all possible values of the fundamental physical constants (e.g. velocity of light c) and all the possible physical laws (e.g. gravitational Newton's law), P is a set of the Universes with the values of the fundamental constants and physical laws which admit the existence of observers (not necessarily the human), and T is the test (observation), which out of the set of all possible universes W described by a certain physical theory selects the Universe which belongs to P.

Some interpretations say that *either* there exists an Intelligent Designer, who planned the Universe in a way which allows humans to exist, *or* that there exists an infinite set of all possible universes (with humans occupying one, or some of them). However, according to Heller (2008), God as Omnipotent could have created both one Universe and also infinitely many Universes (though in many of them there is no life of our type). Then, the hypothesis of many Universes does not exclude the hypothesis of a God (it is even more attractive from the point of view of the power and Omnipotency of God, i.e., theologically)

4. The Multiverse

The many-worlds interpretation of quantum mechanics by H. Everett is very philosophical and it has been challenged strongly by the founders of quantum mechanics (especially those who believe in statistical interpretation and in Copenhagen interpretation) (Barrow and Tipler 1986). However, contemporary superstring theory seem to give a firm framework for considering different scenarios of evolution of the fundamental sets of constants of Nature and the laws of Nature. In superstring theory there are many ways of symmetry breaking and choices of a quantum mechanical vacuum. Since the basic space-time of the superstring theory is 10-dimensional, then a 9-dimensional space (plus time) is compactified into a 3-dimensional world in a couple of hundreds (500) ways (topological cycles). There are about 10 fluxes which can wrap on these topological cycles giving 10^{500} options (or different worlds) fitting exactly the idea of Everett – this is called the Multiverse. It also has its own realization in cosmology by the idea of eternal inflation – each vacuum bubble has its own vacuum with different laws and constants of nature (e.g. Barrow 2002, Pogosian et al. 2004).

Finally, there is the question of whether an idea of the multiverse is testable. In order to answer this question we have to answer the question of what we mean by the multiverse. In fact, there are two possibilities. The multiverse is:

1. The set of pieces of our universe each of them having different physical laws and constants.
2. The set of completely independent entities each of them having different physical laws and constants.

The possibility 1. seems to be testable, while the possibility 2. does not seem to be easily testable.

When testing the multiverse, one would have to first think of challenges:

Challenge 1. Construct an alternative scenario (to a unique we have and live in) of the evolution of the universe which was consistent and allowed life (but not necessarily of 'our type').

Challenge 2. Construct a consistent scenario of the evolution of the universe which would not allow life of any type (this does not seem to be easy because of the fine-tuning/anthropic bounds which determine our way of thinking aimed at a life-contained universe).

By constructing an alternative scenario we mean the construction of all important physical processes which presumably took place in the universe such as: big-bang, inflation, nucleosynthesis, galaxy formation, etc.

Some of us may ask if an idea of the multiverse is practical and not a totally science fiction. This might be viewed as follows (cf. Barrow and Tipler 1986). Classical cosmology (Einstein field equations) is based on only one solution out of an infinite number (in terms of mathematics – an aleph-one which is supposed to be equal to the set of real numbers) which realizes as our universe (or perhaps a piece of it). However, quantum cosmology needs all the classical solutions to be present in the quantum solution (which is a wave function) to get a probability of creating one of the universes (of a certain radius). Then, in classical cosmology, one needs some initial conditions as physical laws to resolve the problem of choosing 'this' solution (a point) which is our universe, and not another solution which is not our universe. In quantum cosmology (which reflects an idea of the multiverse) – all the initial points (universes) are present in the solution, and there is no need for the initial conditions at all, since now all the logically possible universes exist (Kiefer 2007). In that sense we enlarge ontology (all universes instead of one) and reduce the physical laws (no need for initial conditions). This seems to be quite reasonable and, what is more, this enlargement is analogous to the spatial enlargement of the Universe in the Copernican system. Amazingly, such an enlargement was one of the points of a criticism of the Copernican system by opponents who applied Occam's razor. In the case of the multiverse, we enlarge the spatial size of our universe by adding extra universes (pieces of our universe) which have different physical constants and laws and which *are real,* though they may seem to be apparent once we take them classically as one of the realizations of our mentally allowed opportunities which are reflected in the mathematical formulation of some field equations having many different solutions.

5. Remarks

Let us may make a couple of philosophical remarks which refer to the problem of varying physical constants.

Philosophically, the idea of varying fundamental constants is quite attractive since it gives an opportunity for considering all possible options of the evolution of the universe which are logically possible and mathematically admissible.

The universe understood by 'everything which exists' is quite obvious, but it may also be extended to obey 'everything which is thought of' and presumably not real, though if it is applied to the idea of multiverse, then 'everything' contains both – everything which exists, and everything which is thought of.

It is perhaps much easier to test the idea of a multiverse which is understood as a part of our universe with evolving (and so different) physical constants than the multiverse in the sense of decoherent (and so fully separated) histories.

6. Acknowledgments

This work was partially supported by the National Science Center grant No N N202 3269 40 (years 2011-2013).

References

Accetta F.S., Kraus L.M., Romanelli P., "Phys. Lett." 1990, no. B248, p. 146.

Albrecht A., Magueijo J., "Phys. Rev." 1999, no. D59, p. 043516.

Barrow J.D., Tipler F.J., *The Anthropic Cosmological Principle*, Oxford University Press, Oxford 1986.

Barrow J.D., "Phys. Rev." 1999, no. D59, 043515.

Barrow J.D., *The Constants of Nature*, Vintage Books, London 2002.

Barrow J.D., "Phys. Lett." 2003, no. B564, p. 1.

Barrow J.D., "Ann. Phys." (Berlin) 2010, no. 19, p. 202.

Brans C., Dicke R.H., "Phys. Rev." 1961, no. 124, p. 925.

Carter B., [in:] *Confrontation of cosmological theories with observation*, ed. M. Longair, Reidel, Dordrecht 1974.

De Sitter W., "Proc. Royal Acad. Amsterdam" 1917, no. 19, p. 1217.

Dirac P.A.M., "Nature" 1937, no. 139, p. 323.

Dirac P.A.M., "Proc. Roy. Soc. A" 1938, no. 165, p. 189.

Douglas M.R., "JHEP" 2003, no. 0305, p. 046.

Duff M.J., Okun L.B., Veneziano G., "JHEP" 2002, no. 03, p. 023.

Eddington A.S., *The Mathematical Theory of Relativity*, Cambridge University Press, Cambridge 1923.

Eddington A.S., *New Pathways in Science*, Cambridge University Press, Cambridge 1934.

Everett H., "Rev. Mod. Phys." 1957, no. 29, p. 454.

M. Gasperini and G. Veneziano, "Phys. Rept." 2003, no. 373, p. 1.

Gödel K., "Rev. Mod. Phys." 1949, no. 21, p. 447.

Guenther D.B., Krauss L.M., Demarque P., "Astroph. Journ." 1998, no. 498, p. 871.

Horava P., Witten E., "Nucl. Phys." 1996a, no. B460, p. 506.

Horava P., Witten E., "Nucl. Phys." 1996b, no. B475, p. 94.

Heller M., *Podglądanie Wszechświata*, Wydawnictwo Znak, Kraków 2008 – in polish.

Ivanchik A., Rodriguez E., Petitjean P., Yarshalovich D., "Astron. Lett." 2002, no. 28, p. 423.

Johnstone-Stoney G., "Phil. Mag." 1881, no. 5, p. 381.

Jordan P., "Zeit. Phys." 1959, no. 157, p. 112.

Khoury J., Steinhardt P., Turok N., "Phys. Rev. Lett." 2003, no. 91, p. 161301.

Kiefer C., *Quantum Cosmology*, Oxford University Press, Oxford 2007.

Laskar J., Robutel P., "Nature" 1993, no. 361, p. 608.

Magueijo J., Barrow J.D., Sandvik H.B., "Phys. Lett." 2002, no. B549, p. 284.

Mota D.F., Barrow J.D., "Mon. Not. Roy. Astr. Soc." 2004, no. 349, p. 291.

Murphy M.T., Webb J.K., Flaubaum V.V., Prochaska J.X., et al., Mon. "Not. Roy. Astron. Soc." 2001, no. 327, p. 1237.

Olive K.A., Pospelov M., Qian Y.-Z., Manhes G., et al.; arXiv: astro-ph/0309252.

Okun L.B., "Sov. Phys. Usp." 1991, no. 34, p. 818.

Petrov Yu.V., Nazarov A.I., Onegin M.S., Petrov V.Yu., et al., "Phys. Rev." 2006, no. C74, p. 064610.

Pogosian L., Vilenkin A., Tegmark M., "Journ. Cosmol. Astrop. Physics" 2004, no. 0407, p. 005.

Polchinski J., "Rev. Mod. Phys." 1996, no. 68, p. 1245.

Polchinski J., *String Theory*, Cambridge University Press, Cambridge 1998.

Randall L., Sundrum R., "Phys. Rev. Lett." 1990a, no. 83, p. 3370.

Randall L., Sundrum R., "Phys. Rev. Lett." 1999b, no. 83, p. 4690.

Sandvik H. et al., "Phys. Rev. Lett." 2002, no. 88, p. 031302.

Shlyakhter A.I., "Nature" (London) 1976, no. 264, p. 340.

Susskind L., *The anthropic landscape of string theory*, arXiv: hep-th/0302219, 2003.

Uzan J.-P., "Rev. Mod. Phys." 2003, no. 75, p. 403.

Weyl H., "Ann. Phys." 1919, no. 59, p. 129.

Weyl H., "Naturwissenschaften" 1934, no. 22, p. 145.

Williams J.G., Turyshev S.G., Boggs D.B., "Phys. Rev. Lett." 1996, no. 93, p. 26101.

DeWitt B., Neill Graham R., eds., *The Many Worlds Interpretation of Quantum Mechanics*, Princeton University Press, Princeton 1973.

Part III

Ultimate Causality

Bogdan Dembiński
University of Silesia, Katowice

Causality Issues in Ancient Greek Philosophy

'What is visible opens us to what is invisible'
Anaxagoras

The issue of causality is one of the oldest and most serious issues in the history of European thought. One can claim with full conviction that its presence was meaningful to the form assumed by Greek thought. The heart of the matter is the conviction that everything which exists in any way has, and must have, its cause. It seems that such a stance is a common feature of all cultures and it is hard to find a cultural circle in which it would be absent. The difference comes down to the ways in which the question was answered. In the beginning the Greeks, like other peoples, sought justification, primarily in the area of mythology and theology (Theogonists and Cosmogonists). Soon however, the turnabout occurred which changed the character of the philosophical question about the cause, the turnabout which determined the direction of the development in philosophy and science. Greek philosophers became convinced that there was an essential difference between that of cause and the nature of result, and result is not only a simple development of cause. The decisive role was undoubtedly played by Anaximander, Thales' student. The Greek philosophers had a specific 'world physics', as they thought that its form could amount to four states (today we would say phase states): solid, liquid, gaseous and the state of diverse energy. They correspond with the elements: earth, water, air and fire. In the beginning, the Greeks claimed that one of the elements should be regarded as a cause. It is impossible when we distinguish between the nature of cause and result since a cause cannot be its own result.

While asking about the cause of the world, it must be claimed that it might be only something which is unlimited. Anaximander termed it – apeiron[1]. In this manner he expressed the conviction prevalent among Greeks, that the visible Universe came into existence from indeterminacy, the state in which there was no order, and no structure. Others would later confirm this: Pythagoreans would talk about the limitlessness principle, Anaxagoras and Empedocles about the primary mixture, Democritus about the vacuum, Plato about the indeterminate space, Aristotle about the indeterminate primary matter[2]. One

---

[1] All quotations from: *The Presocratic Philosophers. A critical history with a selection of texts*, by G.S. Kirk, J.E. Raven, Cambridge 1957.

Aristotle: "All the physicists make the infinite a property of some other nature (heteran tina physin) belonging to the so-called elements, such as water or air or that which is intermediate between these"; Aristotle, *Phys.* 203a.

Anaximander: "He says that it is neither water nor any other of the so-called elements, but some other apeiron nature, from which come into being all the heavens and the worlds in them"; Simplicius, in *Phys.* 24, 13. "...some other apeiron nature, from which come into being all the heavens and the worlds in them. And the source of coming-to-be for existing things is that into which destruction, too, happens, 'according to necessity; for they pay penalty and retribution to each other for their injustice according to the assessment of Time"; Simplicius, *Phys.* 24, 17.

[2] Philolaos: "All existing things must necessarily be either Limiting, or Non-Limited, or both Limiting and Non-Limited. But they could not be merely Non-Limited (nor merely Limited) Since however it is plain that they are neither wholly from the Limiting nor wholly from the Non-limited, clearly then the Universe and its contents were fitted together from both the Limiting and Non-Limited"; Stobajos, *Anth.* I, 21, 7a.

Empedocles: "Empedocles of Acragas ... holds that the air that was separated off from the original mixture of the elements flowed around in a circle"; Plutarch, *Strom. ap. Eusebium* P.E. I, 8,10.

Anaxagoras: "So they assert that everything is mixed in everything, because they saw everything arising out of everything. But things, as they say, appear different from one another and receive different names according to the nature of the thing that is numerically predominant among the innumerable constituents of the mixture"; Aristotle, *Phys.* A 4, 187a.

Democritus: "Democritus ... calls space by these names: the void, nothing, and the infinite, while each individual atom he calls 'hing' [i.e. 'nothing' without 'not'], the 'compact' and 'being'"; Simplicius, *De celeo* 242, 18.

can see the coincidence with Hesiod's theological thought, his conception of chaos from which the visible Universe emerged. The genius of the Greek conception is in the statement that the reason of the world is limited by its cause (peras). Every being is therefore what it is, owing to its inherent individual limit, which determines its essence and distinguishes it from other beings. Greek philosophers also drew attention to another aspect of the issue. They connected the question about cause and limits with the question about shape and form that indeterminacy would assume.

It is an essential change in the cognition of the essence of cause. There is the conviction that the decisive role in creating the form of the visible world is played by a system of limits, which determinate the cosmic order and harmony (taksis, diakosmenon, kekosmenos, syntaksis, katakosmein). The world structure is the result of an arrangement and connection of constituent elements[3]. Greeks would

Leucippus: "Leucippus holds that the whole is infinite [...] part of it is full and part void. Hence arise innumerable worlds, and are resolved again into these elements. The world come into being as follows: many bodies of all sorts of shapes move, by abscission from the Infinite into a great void; they come together there and produce a single whirl, in which, colliding with one another and revolving in all manner of ways, they begin to separate apart, like to like"; Diogenes Laertios IX, 31.

Plato: "Wherefore, let us not speak of her that is the Mother and Receptacle of this generated world, which is perceptible by sight and all the senses, by the name of earth or air or fire or water, or any aggregates or constituents thereof: rather, if we describe her as a Kind invisible and unshaped, all-receptive, and in some most perplexing and most baffling way partaking of the intelligible (hora aei B.D)"; Plato, *Timaeus* 51a.

Aristotle: "Most of the earliest philosophers conceived only of material principles as underlying all things. That of which all things consist, from which they first come and into which on their destruction they are ultimately resolved, of which the essence persists although modified by its affections – this, they say, is an element and principle of existing things (prote hyle, B.D). Hence they believe that nothing is either generated or destroyed, since this kind of primary entity always persists"; Aristotle, *Metaph.* 983b.

Hesiod: "Verily first of all did Chaos come into being, and then broad-bosomed Gaia [earth], afirm seat of all things forever"; Hesiod, *Theogony* 116.

[3] Plato: "Yes, I believe I understand; I think you mean that the infinite is one class and the finite is another class among existing things; but what you wish to designate as the third class, I do not comprehend very well. And as to the third class, understand that I mean every offspring of these two which comes into being as a result of the measures created by the cooperation of the finite"; Plato, *Philebos*, 26cd.

name the Universe after order – Cosmos. What is it itself then? The Greeks would answer that it is: the organised, *ordered structure of limits* which determine the form assumed by what is limitless. The question about the cause of the world will be transformed into the question about the essence of the cosmic order and will henceforth denote comprehension of the limits which organise this order. They grasp them as forms superimposed on the original, indeterminate material which decide about the ways to shape primary elements. Greek philosophers assume their multiplicity. In their understanding of limits the language of mathematics turns out to be extremely useful, above all the language of geometry (shapes) and the language of arithmetic (numbers and numerical ratios).

What causes such an orderly, limited structure of the world to emerge? The Greeks believed that the decisive role is played by interactions and relations among elements. Those interactions are the *ways to connect* primary elements with one another. So this is about certain proportions of the elements, their arrangements and the locations they take in the indeterminate space. Such interactions can be examined in two ways. Either by means of the philosophical method (dialectic), analysing connections and relations among the notions which describe the elements, or (precisely) in examining interactions by means of various mathematical, geometrical and algebraic methods. In both cases we analyse which connections occur, may occur and which are totally impossible. In Greece it gave rise to the conception, uncommon from the perspective of the history of science that the mathematical natural science conception (phyisikotera ton mathematon), in which understanding of the cosmic order must be strictly connected with its mathematical description. This conception would reach its full development within the walls of the Platonic Academy (in works of Eudoksos of Knidos, Speuzipos, Xenokrates and Herakleidos Pontikos).

The last issue concerns the question about the cause of the existence of limits, their form and multiplicity. Greeks find the explanation in the theory of the Supreme Principles of being which define the argument of ultimate justification. They assume that these principles might be: the One, One and Many, the Opposites, the Limit and the Limitlessness, Love and Hatred, the Cosmic Intellect (Logos), the Divine Mind or the Divine Organizer of the world (Demiurg)[4]. The effect of those principles is therefore a determined form of each limit, as well as their arrangement, which shapes an indeterminate material and decides about the form of the visible world. The principles establish the end of the justification one cannot go beyond.

These problems were precisely codified in Aristotle's philosophy which proposed the classical theory of four causes: material, formal, efficient and final. The material causality answers the question of what something arises from (to hyle, to hypokeimenon). The formal cause defines what is something that arises (to ti en einai). The effi-

---

[4] Parmenides: "Nor is it divisible, since it is all alike; nor is there more here and less there, which would prevent it from cleaving together, but it is all full of what is. So it is all continuous; for what is clings close to what is"; Simpilcius, in *Phys.* 145.

Plato: "...and the ancients, who were better than we and lived nearer the gods, handed down the tradition that all the things which are ever said to exist are sprung from one and many"; Plato, *Philebus*, 16 cd.

Heraclitus: "...is necessary to know that war is common and right is strife and that all things happen by strife and necessity"; Orygenes, *C. Celsum* VI, 42.

Pythagoreans: "Others of this same school hold that there are ten principles, which they enunciate in a series of corresponding pairs: (1) Limit and the Unlimited"; Aristotle, *Metach.* 986a.

Empedocles: "He makes the material elements four in number, fire, air, water and earth, all eternal, but changing in bulk and scarcity through mixture and separation; but his real first principles, which impart motion to these, are Love and Strife"; Simplicius, *Phys.* 25.

Anaxagoras: "And the things that are mingled and separated and divided off, all are known by Mind. And all things that were to be, all things that were but are not now, all things that are now or that shall be, Mind arranged them all, including this rotation in which are now rotating the stars, the sun and moon, the air and the aither that are being separated off"; Simplicius, *Phys.* 164,24.

Plato: "He was good (Demiurgos B.D.), and in him that is good no envy ariseth ever concerning anything; and being devoid of envy He desired that all should be, so far as possible, like unto Himself. This principle, then, we shall be wholly right in accepting from men of wisdom as being above all the supreme originating principle of Becoming and the Cosmos"; Plato, *Timaeus* 29e.

cient cause defines why something arises (he arche tes kineseos). Finally, the final cause answers the question thanks to what something that arises in general arises (to tou eneka, to telos). Aristotle acknowledged those questions as fundamental in constructing any philosophy or knowledge, and he identified wisdom with the cognition of the primary principles and the primary causes[5].

He expressed thereby the essence of the Greek attitude towards the world, the attitude which in examining causes and principles views the most profound foundations to create philosophy and science. It became the basis of European rationality and the source of creating a technological civilization. What is also important is the fact that after almost three thousand years this project has not undergone a major change.

Looking from the perspective of the present-day natural science we can ask: which of the questions asked by the Greeks should be recognized as essential for the contemporary debate on the nature of the world?

I think that it still remains the question about cause, but focused precisely on the issue of the essence of the cosmic order (astrophysics and cosmology). Today it has been broadened to the order of the micro world. This order (both explicit and implicit) should be uncovered. Extracting what is implicit, from behind the cover of what is explicit, Greeks called uncovering the truth, and determined with the notion – aletheia. To uncover the truth is to reveal the implicit and explicit cosmic order. Only the intellect can accomplish it and it does so, because it is itself a part of this order and its structure is created according to the same principles and regularities as the whole Cosmos is created (Plato termed it anamnesis). The aim of the action (the essence of science) is cognition of this structure, thus the cognition of the whole system of con-

[5] Aristotle: "The reason for our present discussion is that it is generally assumed that what is called Wisdom is concerned with the primary causes and principles"; Aristotle, *Metaph.* 981b.

nections and relations created by constituent elements. Therefore the cognition must aim on the one hand at cognition of the very elements and, on the other hand, (first of all) at examining the system of connections which occur among them. This examination is searching for 'harmony', 'inner schema', 'cosmic pattern', according to which the world order was created. We are still doing it at it was done in Antiquity by means of intellectual models. Both then and now they are mostly mathematical models. In this manner we have take an roundabout route and, like the Greeks, we try to read the order of Cosmos by referring to its mathematical description. And it does not seem to be a specific kind of the miracle mentioned by Eugene Wigner. One should rather claim that the true miracle would be the lack of such ability. Then nothing could be comprehended and nothing could be cognized. So, the essence of cause must be searching for the essence of cosmic order. If we understand the Pithagorean notion of harmony as a notion of cosmic order, we can quote the great Philolaos:

'This is how it is with Nature and Harmony: the Being of things is eternal, and Nature itself requires divine and not human intelligence; moreover, it would be impossible for any existing thing to be even recognised by us if there did not exist the basic Being of the things from which the Universe was composed, namely both the Limiting and Not-Limited. But since these Elements exist as unlike and unrelated, it would clearly be impossible for a Universe to be created with them unless a harmony was added, in witch way (harmony) did come into being. Now the things which were unlike and unrelated and unequally are necessarily fastened together by such a harmony, through which they are destined to endure in the Universe'[6].

[6] Philolaos. Fr. 6. Stobajos. Auth. I, 21.

William R. Stoeger, S.J.
Vatican Observatory
Steward Observatory
The University of Arizona

Cosmology, Evolution, Causality and Creation: The Limits, Compatibility and Cooperation of Scientific and Philosophical Methodologies

Over the past century there has been an explosion of knowledge and understanding about all aspects of nature, and of the vast universe of which we are a part. Along with the emerging details of the physics, chemistry, biochemistry, and biology from focused scientific research has been a more refined awareness of the many different intricately related factors, or 'causes', which are at work in nature and in the universe. This has been particularly evident as we have explored the emergence of new and more complex systems from simpler ones, and their multi-layered operations, within a given environment, culminating in living organisms manifesting self-consciousness and learning. Describing many of these factors which contribute to determining what happens and what emerges in the process of complexification is certainly what the natural sciences are all about. But there are some of them which are immune to, or only partially vulnerable to, scientific investigation. And so it falls to philosophy to reflect upon all these causes – whether subject to scientific investigation or not, to characterize and categorize them, and to explore their inter-relationships and operation.

In this paper we shall first briefly describe what we know about the make-up and the history of the universe. Secondly, I shall discuss the order, the processes, the relationships and the regularities – the laws of nature – which have been and are essential to its evolution and operation, and to the emergence and operation of all the systems which arise within it – including life and consciousness. As a natural extension of these considerations we shall, thirdly, discuss the different types and models of causality, and their relationships with one another. Finally, after briefly speaking of what we know of the Planck era and quantum cosmology, we shall explore a rather different kind of 'causality', which is clearly a consideration that takes us beyond the natural sciences into the realm of philosophy. This is the causality – which is often called 'primary causality' – which many postulate and hold as the ultimate source of existence and order. It often is described under the rubric of 'divine creation'. Strictly speaking it is a philosophical idea that was taken over by the monotheistic religions in order to formulate the role of a divine creator and specify how the Creator's action interfaces with the natural causal fabric we come to know and understand from scientific investigation. In a very definite sense, it is 'causality beyond causality', and a model for the ultimate ground of all other causes, and of existence itself.

1. The Structure and History of the Universe

We know that our observable universe is immense, extending of the order of 40 billion light years in every direction, and consisting of hundreds of billions of galaxies, each galaxy harboring as many as 200 or 300 hundred billion stars. On at least one planet around one of those stars in one of those 100's of billions of galaxies, extremely complex systems and life emerged, eventually evolving organisms with the capability of self-consciousness, abstract thinking, symbolic communication and learning, and advanced technological culture.

This development depends first of all upon the early emergence of hierarchical structure in the universe – planets, asteroids and comets around stars, clusters of stars, galaxies, clusters of galaxies, superclusters and voids. None of these were present at the very beginning, shortly after the Big Bang. Furthermore, that macroscopic structure – more particularly the advent of stars – was essential for the emergence of microscopic structure, chemistry. It is through stellar fusion processes that all the 92 natural elements were produced, with the exception of hydrogen, helium and a little bit of lithium. Without stars we would have no chemistry to speak of, and no complex systems, and no life or consciousness.

There is, however, a remarkable and puzzling fact about the contents of our universe that is not at all apparent in this visible and well-known dual organization of the cosmos. All the visible mass in the universe, consisting of anything made up of atoms – protons and neutrons (baryons) and electrons (leptons) – accounts for only about 4% of its total mass energy. About 23% is made up of non-baryonic, weakly interacting dark matter, and another 73% of what we call dark energy, probably vacuum energy, which is driving the universe's accelerated expansion. Much less than 1% at present is due to all the radiation which fills the universe[1].

So, we can summarize the principal characteristics of the universe today as being 13.7 billion years old (since the Big Bang, or better its emergence from the Planck era), extraordinarily vast with over 100 billion galaxies, expanding, cooling and gently accelerating, evolving on all scales, being very smooth on large scales and very lumpy on small and intermediate scales, having much more dark matter than visible baryonic matter, with a relatively large amount of dark energy, and being fine-tuned for complexity and bio-friendly.

Before sketching the history of the universe, it is helpful to recall that its basic evolution can very simply be described as continual expansion from hotter, smoother and more simple configurations to ever cooler, more lumpy and more complex ones. As the universe cools and becomes lumpy, new things become possible and emerge.

[1] There are an enormous number of references to the cosmological details we summarize here. A few of these worth checking are Ellis (1993), Liddle and Lyth (2000), Rees (2001), and Hobson, Efstathiou and Lasenby (2006). The first chapter or two of both Kolb and Turner (1990), Dodelson (2003) and Weinberg (2008) are also recommended.

Cosmic evolution thus sets the stage for complexity and life. We need a cool, highly structured, complex and richly differentiated environment for life to be possible.

This expansion and cooling began with the emergence of the universe from what is now known as the Planck era (the Big Bang), when the universe was so hot that not even space and time, gravity or the other physical forces existed as we know them. We are unsure about the physics that describes such an extreme state – though many physicists and mathematicians are exploring the possibilities, in a field that is called 'quantum cosmology'. This emergence from the Planck era happened about 13.7 billion years ago, as I have already mentioned. Within the first second after that, what is called inflation (a very brief period of extremely rapid expansion and cooling), followed by the exit from inflation and cosmic re-heating occurred. A couple of minutes later the various isotopes of hydrogen and helium were synthesized, along with a little bit of lithium, the lightest metal.

During the following 300,000 years, the universe continued to expand and cool slowly as a very smooth (homogeneous) mixture of matter and radiation. The baryonic matter was ionized and thus was strongly coupled to the radiation via electron scattering. This prevented it from clumping – although the dark matter was already beginning to do so. Once the visible (baryonic) matter cooled enough so that most the protons captured electrons, becoming neutral, the dominant scattering of radiation by free electrons ceased, and radiation decoupled from the matter. The over-dense regions of matter then could clump under the influence of gravity to begin to form clusters of galaxies, galaxies and stars. At that same time, the universe became transparent to radiation, and the cosmic microwave background photons were released to propagate throughout the universe with very little further interaction with free electrons.

In the succeeding billions of years, galaxies and stars formed and spewed chemically enriched material into the cosmos, enabling chemical evolution and possibly biological evolution on planets and moons. After about 9 billion years of cosmic evolution, our Solar System formed (4.6 billion years ago) – the Earth forming along with the Sun and other planets, in the collapse of the pre-solar nebula. Life itself originated about 4.0 billion years ago on the Earth, and the earliest prokaryotes go back at least 3.5 billion years. About 2.5 billion years ago free oxygen accumulated in the atmosphere, as the waste product of anaerobic bacterial metabolism. The earliest eukaryotes emerged 1.5 billion years ago, the earliest animals 700 million years ago, the earliest hominids 4-5 million years ago, and *Homo sapiens* 400 thousand years ago. This run-down summarizes in very broad outline the major transitions in the history of the universe and of the earth. Obviously, since the Solar System developed so late, relatively speaking, in cosmic history, it is certainly possible that life and consciousness emerged much earlier elsewhere in the universe. But we just don't have any evidence of that. But, as people often say, absence of evidence is *not* evidence of absence – particularly in cases like this!

2. Cosmic Order, Processes and Relationships

The central focus of our discussion now comes to the fore: What causes or factors contributed to our universe being how it is, and evolving as it has evolved? What combinations of determinants at the Planck-era 'beginning', or before that, and during its evolution have effected the emergence of the different stages of cosmic organization and their characteristics, and those of the many different levels of systems within each of them? We generally lump all these factors or causes under the overarching concept of the cosmic order, or the 'laws of nature'. It is that order which guides the dynamics of the universe and of nature itself in cosmic evolution, chemical evolution and biological evolution, leading to the emergence of new and more complicated systems, and the dominance of new global and local environments. If we reflect more fully on what we know about nature at all its stages and at every layer of complexity, it is very clear that what is of central importance are the many levels of dynamic relationships among basic entities as well as among more complex systems. Often there are complex networks of innumerable relationships intricately guiding the maintenance and functioning of a given system, and of that particular system in concert with other systems within a larger system, and so on. This involves certain regularities of function and is both the product and the promoter of an ongoing evolving intrinsic order and development. In the course of that evolution, there is always differentiation of entities and systems, and of the relationships among them, together with specific combinations, leading to new emerging structures and relationships. And so, for example, we have protons, neutrons and electrons forming the atomic nuclei, and atoms, of the different elements, under the operation of the strong, weak and electromagnetic interactions, which effect their relationships with one another. Under electromagnetic relationships among the atoms supporting the laws of chemical bonding, we have the almost innumerable combinations of inorganic and organic compounds and crystals. These in turn become the building blocks of larger molecules, each with very specific functionalities and characteristics, which are not reducible to their components.

As I indirectly hinted just now, the most basic – or almost basic – ways of relating at the level of physics are given by the four fundamental forces or interactions: gravity, the electromagnetism, and the strong and weak nuclear forces. At very high energies and temperatures there is evidence that these four forces successively unify – become identical in strength and character. We suspect that the physics of the Planck era was governed by a single unified 'superforce'. Working forwards in time, that superforce first split into gravity and the grand unified theory (GUT) force, as the universe exited from the Planck era. At significantly lower energies this GUT interaction split apart into the strong nuclear force and the electroweak force[2]. And

[2] Again, there is a very large literature describing what we know about the very early universe. For a brief introductory account, see the relevant summary sections of Kolb and Turner (1990).

then at relatively low energies the electroweak force splits into electromagnetism and the weak nuclear interaction. This primordial differentiation of relationships also involves and was accompanied by the emergence or condensation of various fundamental particles and combinations of fundamental particles. All of this occurred within the first second after the exit from the Planck era – as the temperature plummeted rapidly after inflation and reheating. It is these fundamental relationships and particles which form the most fundamental and basic level of organization in reality. At slightly more advanced and complex levels we have particles combining through these relationships. This results in a large variety of new systems with completely new properties and capabilities, along with more diversified relationships – e. g. the laws of chemical bonding. These new systems and relationships have the underlying fundamental components and relationships as necessary conditions, but they and their relationships and properties are not completely reducible to them, particularly when they become bearers of information about their environment and its history at the level of biochemistry and molecular biology.

When we talk about these relationships, and many others at higher levels of complexity – the laws of nature, broadly speaking – I should specify what I include in them. What I mean by 'the laws of nature' are all the relationships, processes and regularities inherent in the natural world – all those which constitute the order in nature and in the universe. In saying this, it is crucial, I believe, to distinguish between: 1) *Our* laws of nature; and 2) the laws of nature as they actually function (Bunge 1979, pp. 249 f; Stoeger 1993). The latter are all the relationships, processes and regularities that enable and characterize nature to be what it is, whether we have recognized them, understood them, or modeled them with reasonable accuracy, or not. *Our* laws of nature, on the other hand, are the relationships, processes and regularities we have recognized, described and modeled with some success, and understood, at least to some degree. It is clear that there are relationships and regularities we have not recognized or modeled with any success, or which the natural sciences are not competent to investigate. This is important. For there are certainly phenomena or experiences which fall outside the laws of nature as we have understood and modeled them, but involve relationships or aspects of reality which actually function. This is particularly true in cases where we have evidence that something we can't explain actually occurred.

At this point, following on our discussion of order within nature and the universe, it is helpful to stand back and ask ourselves what are some of the general outstanding characteristics of reality as we have come to know and understand it through the natural sciences, and our reflection upon that knowledge (Stoeger 2002, 2008) You may have your own list, but I think it is important to take note of them. On my list are the following: Order and intelligibility, locality within a global context, highly differentiated and dynamic relationality, many nested levels of complexity, evolution on all scales, continual emergence of new systems and relationships, the unity and solidarity of nature through common physics and chemistry, formational and functional integrity (nothing outside of nature is needed), contingency and the important role of chance, transience and fragility, some definite directionality, the importance of time, and the apparent fine-tuned character for complexity it exhibits.

We cannot spend time describing all of these, but some further discussion of some of them is appropriate. We have already highlighted relationality. Reality is the way it is, because of the extraordinary organization of components and systems into networks, and networks of networks, established by intricately coordinated relationships among their components and sub-systems at every level. It is these relationships that enable the multi-layered nested, or hierarchical, levels of complexity that are constitutive of reality. The formational and functional integrity of nature essentially means that nature is complete in itself (Van Till 1986) – that in the emergence and evolution of the universe and of the systems within it, nothing outside of what is given in nature is needed to explain how new systems and entities arise from more basic entities and systems, or how these systems function. Nature and evolution are not micro-managed from the 'outside'.

Along with that integrity is what we might call the unity and solidarity the universe and nature at large exhibits. Everything in the universe is connected in some basic way with everything else. Most strikingly, everything in our observable universe traces back to the Big Bang – the same Planck era. Furthermore, during the history and development of the universe, the components and fundamental relationships of everything in it originated and developed together, guided by the same laws of nature – the same physics and chemistry. This is true also of everything in our galaxy, our Solar System and our planet. Though it is very probable that the biologies that may have developed independently in other stellar systems are very different from the one which has developed on Earth, it is clear that they would be supported by the same basic physics and chemistry – but undoubtedly with somewhat different molecular building blocks.

Balancing this global or cosmic unity and solidarity has been the gradual separation of local, virtually independent regions and systems at many different levels, relative to outside influences. This was essential for those systems to develop and diversify further into more and more complex networks of systems, with ever richer repertoire of capabilities. Locality protects systems from distant disruptive and damaging influences, while enabling some limited, well controlled and sustaining interaction with the surrounding environments.

And then there is role of chance or contingency throughout nature. Everything depends on something else, and often on factors which are afflicted by some randomness. This is due to the extremely complex connections and interactions things have with one another, and also to quantum indeterminacy in the fundamental structure of matter. That has its definite limits, but is very real and influential nevertheless. There are two important things to notice about this characteristic of nature. The first is that what we refer to as chance events always occur within a larger framework of order and relationship. Secondly, chance, and the resulting flexibility and openness of nature, is absolutely essential for the emergence and development of novel systems and structures. If there were no chance events, the resulting rigidity would stifle any further evolution or complexification. Finally, there is the transience and fragility of all systems in the universe. Nothing lasts forever, except perhaps the universe itself. This important property is absolutely essential for further evolution and emergence of systems, organism and persons. If everything that emerged were permanent, evolution would come to halt. There would be so many bottlenecks that the processes essential to further emergence and development would quickly stagnate (Stoeger 2009).

These general characteristics of reality as we know it through the natural sciences help us appreciate more fully how the universe and the systems within it have evolved into what they are now. As we have already briefly discussed, it is through the intrinsic, evolving order within the universe, embodied in the entities and systems which constitute reality and the relationships, regularities and processes which connect them that evolution and emergence occur. At any given point in time, there are definite outcomes of the operation of those relationships and processes, including some freedom given by 'chance'. Further operation of those, and of the laws of nature deriving from those emergent outcomes often leads to new structures, relationships, processes and regularities – to richer organization.

In every situation the context or environment in which these systems function is always very important. Physical contexts (temperature, density and chemical composition) establish the necessary conditions for the specific operation of the interactions and relationships given by the laws of nature. Thus, because this relationship with context is so important, the outcomes which emerge at any stage are at least partially dependent on the context in which they occur, and reflect or provide information about that context. Furthermore, as is obvious, as evolution and development proceed, the physical, chemical, biological, psycho/sociological context also evolves, under the influence of both local and more global relationships and processes. Again the key general pattern involves the differentiation of components, and the relationships and processes associated with them. Combinations brought about by those relationships lead to further differentiation and complexification, along with novel properties, relationships and capabilities. This evolution continues through further combinations, differentiations, and systemizations, most often leading to complex interactive networks. This results in what we have already described as the hierarchical, or deeply inter-nested dynamical structuring of reality. Moreover, whenever a viable new system emerges, that emergence always involves individuation, separation from its context, while remaining in continual mutual interaction with it.

From this discussion we can briefly summarize the conditions for the development of complexity in our universe. They include locality; differentiation, separation from context, and condensation; openness of the system to its context; dynamic combination through relationships; and emergent systems and behaviors. To these we can add several properties we have not discussed – non-linearity; bottom-up, same-level, and top-down causality; feedback control; and goal-like (teleonomic) behavior.

When we reach the level of living systems, there are additional conditions which are essential: systematic unity and persistence of identity and behavior; self-maintenance within its environment; generation, selection, preservation and effective coordinated implementation of information; modularity and redundancy; and reproduction (Küppers 1990; Ellis 2006). All these insure biological evolution, as long as there are mechanisms which introduce genetic (informational) variation in the living systems involved.

3. Understanding Causation and the Variety of Causes

So far we have sketched in very general terms how our universe has gradually evolved from a very simple configuration just after the Planck era to the universe we see and are part of today. And we have some idea now of the factors and influences which worked together to guide, or perhaps even determine, what has gradually emerged. We often refer to these factors as 'causes'. 'Causes' – what are they? As all of us already know, there are many rather different understandings of 'cause'. Some limit it very specifically to what we often refer to as 'efficient causality'. But many others apply the term more broadly. For the purposes of our considerations here, I opt for a broad concept of 'cause'. Here 'causes' are all those factors which enable the existence, formation, persistence and functioning of an entity, system, organism – whatever contributes to determining a state of affairs. This is an adaptation of Mario Bunge's discussion of causality (Bunge 1979). However, interestingly enough he restricts the use of the word 'cause' to mean 'efficient causality'. But, at the same time, he clearly says. '... several categories of determination contribute to the production of every real event' (Bunge 1979, p. 280). In speaking about the causal universe, we really want to honor, acknowledge and understand more fully all those categories of determination and how they interact with each other. And so we adopt the broader characterization of cause we have suggested above.

Of course, this is just in keeping with the very traditional philosophical classification of causes set forth by Aristotle. As we all know, he spoke of four general types of cause: material causes, formal causes, efficient causes and final causes (cf. Scott 2007). Here we shall refine and elaborate this classification using rather different – and more scientifically derived – categories. Before doing so, it is important to recognize that we can really effectively and clearly analyze and isolate causes only in very simple or ideal situations. In most concrete cases, the actual causal situation is very complex. Usually there are multiple interacting networks, each network itself a dynamic and intricate system of many different closely communicating processes, relationships, and systems. All the relationships mutually modulate and constrain one another in very precise and carefully monitored ways. We would not exist – nor would any other living thing – without that complex dynamical structuring.

So, what types of causes can we point to as we reflect on our scientific understanding of nature? First of all, there are various entities and structures – like particles and fields at the fundamental level of physics. These are reminiscent of Aristotle's material cause. Then there is the differentiation of these entities and structures, which might be as simple as the way a collapsing gas cloud fragments into smaller collapsing clouds in the formation of a cluster of stars – or as complicated as the formation of proto-cells in a chemically rich environment. Related to this are the various processes of interaction and combination, which involve a variety of different types of relationships. There are internal constitutive relationships which hold systems together and enable them to function effectively – in very simple, or in very complex ways. There also are very important external constitutive relationships systems have with their ancestral systems, from which they have inherited key characteristics and capabilities. These are obviously diachronic, but absolutely essential. There are important synchronic relationships of systems with their environment – essential for their maintenance and survivability. Some of the internal organizational and external relationships might be called 'constraints' – they channel and limit the operation of other more efficiently causal influences. But they all are due to organizing relationships of one form or another. The general category of 'relationship' is much broader and more fundamental than that of 'constraint'. The most important relationships for our considerations arise in the development of the systems and organisms themselves – not from the outside – and serve in a highly differentiated way to dynamically connect simpler systems into more complex and functionally distinctive systems. All these relational or organizational causes are really contributors to what Aristotle refers to a 'the formal cause'.

Many causes can also be categorized as bottom-up, same-level or top-down – or, similarly, whole-part, or part-whole, relationships. (Peacocke 1990; Ellis 2006; Auletta, et al. 2008; Ellis 2012) As indicated, bottom-up causes provide the basic underlying necessary foundation for the existence and operation of more complex systems – for instance the basic physical and chemical processes that are essential for the healthy functioning of a human body. Top-down causes often act as organizational constraints on the lower-level causes, marshaling them into effective service to the larger system of which they are a part. Same-level causes are those which effect processes, interactions and relationships on a given level of complexity for the proper organization and function of the entire system. Similarly, for instance in many cases – particularly in biological systems, parts only emerge and function within the framework of the whole. Parts derive from wholes in these systems. But, at the same time, parts contribute essentially to the existence and proper functioning of the whole.

There is also the emergence of a certain teleology (final causality), which stems from the differentiation and therefore the specialization of subsystems within a larger complex system. As the larger system evolves or develops, certain parts of it develop their own internal structures and specific capabilities. Those capabilities serve the larger system in essential ways. And the very reason why that substructure forms, persists and thrives within the system is precisely because it provides an essential contribution or process to the existence and proper functioning of the larger system. Good examples are the heart, or the lungs, of an animal. This is sometime referred as functional teleology (Stoeger 2011), or 'consequence etiology (Depew 2011)'. The very reason why the that substructure emerged is because of the contribution it makes the larger system.

Finally, there is what we have referred to as 'chance'. In many cases, particularly in a macroscopic context, 'chance' is simply a particular influential relationship or interaction with the environment, or something in the environment which disturbs, changes or varies the system in a definite unusual or unforeseen way. But it is perfectly understandable, once we take that 'rogue' influence into account.

In a number of other cases it results from quantum uncertainty – for instance a random genetic mutation – and the way it gets amplified into macroscopic consequences. We recall that it is primarily due to the operation of chance that the universe and nature have the flexibility and openness that is needed to explore all the possibilities leading to the emergence and development of new and more advanced complex systems.

The importance of all these causal factors, including chance, which we often describe generally as the laws of nature, is strongly emphasized, as well as summarized, by Paul Davies, when he says very elegantly and perceptively: '... [the development of complexity] depends on the kind of radical chance permitted by, yet transcending the determination of these very laws [of nature]. ... these laws are not only valid but essential. ... intrinsic creativity is entirely the result of natural processes which are infused with the inherent power of self-organization based on, though not reducible to, these very specific laws of nature' (Davies 1998).

4. Causality Beyond Causality: Creation and the Source of Existence and Order

In this final section we move beyond where the natural sciences can reliably proceed, to consider how we might answer the ultimate causal question: What is the source or 'the cause' of existence and of this cosmic order we have been investigating? But before doing so, we set the stage for our discussion by looking more closely at the Planck era, from which in the so-called Big Bang the universe began to expand and cool. This will help us appreciate some of the limits of physics.

The Planck era, as we have very briefly mentioned above, is the earliest known epoch of cosmic history, characterized by the breakdown of all standard physics, including Einstein's theory of gravitation, general relativity. The universe at that very, very early time was simply too hot for space-time and gravity as we know them to exist. An adequate description of space and time, together with gravity, must rely on a quantum theory of gravity. No one has yet constructed and confirmed one that is reliable or adequate. But many people are working on doing so, developing various types of promising superstring, loop-quantum-gravity, and non-commutative-geometry proposals (Smolin 2001 and references therein). It is likely that any such adequate theory will also involve the complete unification of the four fundamental physical interactions (gravity, electromagnetism, and the weak and strong nuclear interactions). In four-dimensional space-time one enters the Planck era when the temperature of the universe rises above 10^{32} K, well out of the reach of any conceivable particle accelerator. An adequate and reliable theory of the Planck era would tell us how our universe began expanding and cooling, how Planckian physics meshes with classical physics at lower energies – in particular how gravity, and space-time emerged from the primordial Planckian quantum configuration – and possibly what gave rise the Planckian era itself, or what preceded it (Stoeger 2010).

But one thing is clear from these fascinating proposals, and from the limited progress researchers have made so far: Any reliable and adequate theory of the Planck era and the Big Bang requires some physical entity, structure or state along with the detailed physics describing it and its behavior – whatever underlies and explains the Planck era itself. Any such account will always demand, in turn, some further explanation or physical foundation – and ultimately some metaphysical ground or source other than itself and its physical antecedents (Stoeger 2010). A number of quantum cosmologists and philosophers of quantum cosmology maintain that it can tell us how something emerges from nothing. But this is a serious misunderstanding. The 'nothing' here is not simply a vacuum state, but rather nothing at all – no physics, no potential, no space or time, no quantum state, no nothing. Physics simply will never be able to describe an actual physical transition from absolutely nothing to something – something that has existence and is subject to a definite kind of order. There is no physics of absolutely nothing. Physics presupposes existence and order and then goes out and explores how to describe it, including how what exists changes under the influence of the order or the laws of nature to which it is subject. Thus quantum cosmology, or any area of physics or natural science, cannot tell us: 1) why and how something exists rather than nothing at all, and 2) why and how there is order – the ultimate source of the laws of physics – and why and how there is this kind of order rather than some other kind.

It is really philosophy, and by extension philosophical theology, which proposes an explanation how there is something rather than nothing, and how that something is ordered. It does this through the idea or model of divine creation – particularly the version which is often referred to as *creatio ex nihilo,* creation from nothing.

What does divine creation do? Properly understood, philosophically and theologically, it simply and in a very unadorned way provides an ultimate philosophical explanation for the existence and order of reality – of all that the natural sciences explore. It does this simply by proposing a self-subsisting, self-explanatory 'cause' – the Creator – which is the fundamental source of being and order, and in whose existence all existing things participate, in a very limited way (see Stoeger 2010 and references therein). The reasoning behind this basic proposal is that we can clearly argue that from absolutely nothing nothing emerges. So, since there are things that exist, they must come from something else. Obviously they do! But also, since nothing we can point to in our direct experience is self-subsisting – does not account for its own existence and order and does not necessarily exist, and therefore depends on something else for its existence – there must be an ultimate self-subsisting source of being and order. Otherwise, it is impossible to account ultimately for existence or for order. Thus, this Creator is not another entity or process in the universe – it is not scientifically accessible and is not a substitute for physical causes or entities. But it is that which endows – or empowers – them all with existence, order, efficacy, and their own causal capabilities. Thus, it does not claim to provide an alternative to – or substitute for – the causes and explanations quantum cosmology, physics or the other natural sciences propose, describe and investigate.

In a sense physics and cosmology raise the philosophical question of the ultimate source of existence and order, which they themselves cannot answer. This is one of the limitations of the natural sciences. As a result, creation as a philosophical and theological insight or scenario complements – and is compatible with – physics and cosmology, as long as it is properly understood! Furthermore, this philosophical model has been a fruitful philosophical stance in the long-term. It fits in with all other knowledge we have, and has provided a basis for a enlightening philosophical and theological reflection and discussion. In saying this, it is important to emphasize that this fruitfulness in not primarily within the context of the natural sciences, but rather within the context of philosophical and theological investigation. However, indirectly, it does help us to appreciate the limitations of the natural sciences, just as our knowledge and understanding of the natural sciences have helped us realize the limitations of philosophy and theology – and the need for philosophers and theologians to accept and reflect upon the legitimate conclusions of the natural sciences.

There are a number of further precisions that are important to make about divine creation – or, as it is sometimes referred to, God's universal creative action. The first has to do with how we are to understand this Creator as ultimate origin. The Creator and creation is *not* primarily or even necessarily the *temporal* origin of the universe, or of reality (Stoeger 2010 and references therein). The essence of divine creation is that God is the *ontological* origin of all that is – everything is ultimately dependent on God for existence and order. That does *not* imply a temporal origin – though there might have been one. Reality as we know it may have always existed in some form, but from all eternity it would have depended for its existence and order on the Creator, its ontological origin.

Thus, secondly, creation is *not* about an event, but rather about an ongoing relationship of ultimate dependence. The Creator sustains or conserves nature and the universe in its existence and order. Furthermore, it does not as such involve change. Change has to do with created causes and their effects.

Thirdly, it is important to repeat that creation is about the Creator enabling and empowering nature and all things within nature to be and to act on their own as they are, not filling gaps in its causal fabric – except the gap between absolutely nothing and something, and between no order and a specific type of order. Thus, this philosophical proposal satisfies and supports the the formational and functional integrity of creation we emphasized earlier in our discussion. God works within and through nature and the laws of nature, not in parallel. As St. Thomas Aquinas says, the Creator gives to creatures the dignity of being causes in their own right (see Tracy 2013, this volume, and references therein).

There is a fourth point that is also important to consider: The Creator's relationship with each entity or system, though endowing each with existence and order, is different. God's universal creative relationship is highly differentiated. That's simply because each entity or system is different – has a distinct individuality and different capabilities. Another way of putting this is that the Creator 'knows' each thing in its particularity or individuality, not as an instance of an abstract concept. Therefore the Creator supports it in existence as it is in itself – but also at the same time as it is in relationship with other things, that is in its order relative to the rest of nature. The Creator is therefore transcendent – above and beyond all other causes and systems, and also without barrier, obstruction or limitation to His/Her presence, action or relationship. This lack of restriction enables the Creator's immanence – active intimate presence as Creator in nature, and in each segment of nature. But that presence and action is *not* just another cause, process, relationship like the causes, processes and relationships we discern and can study directly using the natural sciences. Rather it is that which supports and maintains everything in its existence and ordered behavior.

Finally, it is crucial to recognize that this creative 'causality' and 'action' can be attributed to the Creator only analogically. God's creative action is radically and completely different from any other kind of action or causality. That's why is it often referred to as 'primary causality', with all other causes – created causes – as 'secondary causes'. The one common characteristic primary causality shares with secondary causality is that it is effective. The difference it makes is simply that without it nothing would exist.

Philosophers and theologians continually stress that God, the Creator, and the creative act/relationship itself cannot be anywhere near adequately understood or comprehended. However, our full range of experience – and our reflection on that experience – discloses something beyond the horizon of secondary causes which transcends them, embraces them and ultimately enables their realization. Our experiences of beauty, of the intricate unity of life and all that is, of meaning and value beyond transience and mortality, of deep love and faithfulness, of heroism and self-sacrifice point to something profound and fundamental beyond what we can probe and comprehend. So does the inexhaustible depth and richness of reality. We cannot grasp what the full range of our experience discloses or points to, but yet we know that there is something there from which everything derives. We sometimes call that Mystery. We can gradually come to understand something about reality, and we are always growing in our appreciation and understanding – through the natural sciences, through philosophy, theology poetry, art, and our most treasured life experiences. But we can never arrive at fully adequate answers or explanations – or anything approaching that. We are always searching for the *least inadequate* accounts of what we experience, observe and study.

5. Conclusion

In our discussion we have reflected upon and discussed the various types of causes, or determining factors, which function within our universe, and within nature – and to some extent how they are related to one another. We have only been able to scratch the surface of understanding and point in worthwhile directions. Many, many other scientists and philosophers have reflected much more fully and profoundly on these causes. Our attempt here has been simply to provide an integrated sketch. We have also spent some time carefully describing the idea of divine creative causality, which transcends the causes which the natural sciences are capable of exploring. We have done this both to indicate the necessity, and support the plausibility, of such transcendent 'causality' – as the ultimate source of existence and order – and also to show that, properly understood, it crucially complements the causal repertoire which the sciences model, and cannot be in conflict with their authenticated findings.

References

Auletta G., Ellis G.F.R., Jaeger L., *Top-down causation by information control: From a philosophical problem to a scientific research program*, "J. R. Soc. Interface" 2008, no. 5, pp. 1159–1172.

Bunge M., *Causality and Modern Science*, Dover Publications, New York 1979, 3rd Edition.

Davies P., *Teleology without teleology: Purpose through emergent complexity*, [in:] *Evolutionary Biology: Scientifiic Perspectives on Divine Action*, eds. R.J. Russell, W.R. Stoeger, S.J., F.J. Ayala, Vatican Observatory Publications, and the Center for Theology and the Natural Sciences, Vatican City and Berkeley, CA 1998, pp. 151–162.

Depew D.J., *Accident, adaptation, and teleology in Aristotle, Empedocles, and darwinism*, [in:] *Biological Evolution: Facts and Theories – a Critical Appraisal 150 Years after the "Origin of Species"*, eds. G. Auletta, M. LeClerc, R.A. Martinez, Gregorian and Biblical Press, Rome 2011, pp. 46–478.

Dodelson S., *Modern Cosmology*, Academic Press, New York 2003.

Ellis G.F.R., *Before the Beginning: Cosmology Explained*, Boyars/Bowerdean, London 1993.

Ellis G.F.R., *On the nature of emergent reality*, [in:] *The Re-emergence of Emergence: The Emergentist Hypothesis from Science to Religion*, eds. P. Clayton, P. Davies, Oxford University Press, Oxford 2006, pp. 79–107.

Ellis G.F.R., *Top-down causation and emergence: Some comments on mechanisms*, "Interface Focus" 2012, no. 2, pp. 126–140.

Hobson M.P., Efstathiou G., Lasenby A.N., *General Relativity: An Introduction for Physicists*, Cambridge University Press, Cambridge 2006.

Kolb E.W., Turner M.S., *The Early Universe*, Addison-Wesley, Redwood City, CA 1990.

Küppers B.-O., *Information and the Origin of Life*, The MIT Press, Cambridge, MA 1990.

Liddle A.R., Lyth D.H., *Cosmological Inflation and Large-Scale Structure*, Cambridge University Press, Cambridge 2000.

Peacocke A.R., *Theology for a Scientific Age: Being and Becoming – Natural and Divine*, Basil Blackwell, Oxford 1990.

Rees M.J., *Just Six Numbers: The Deep Forces that Shape the Universe*, Basic Books, New York 2001.

Scott A., *Nonlinear science and the cognitive hierarchy*, [in:] *Evolution and Emergence: Systems, Organisms, Persons*, eds. N. Murphy, W.R. Stoeger, S.J., Oxford University Press, Oxford 2007, pp. 173–197.

Smolin L., *Three Roads to Quantum Gravity*, Basic Books, New York 2001.

Stoeger W.R., *Contemporary physics and the ontological status of the laws of nature*, [in:] *Quantum Cosmology and the Laws of Nature: Scientific Perspectives on Divine Action*, eds. R.J. Russell, N. Murphy, C.J. Isham, Vatican Observatory Publications and the Center for Theology and the Natural Sciences, Vatican City and Berkeley, CA 1993, pp. 209–234.

Stoeger W.R., *Science, the laws of nature and divine action*, [in:] *Interdisciplinary Perspectives on Cosmology and Biological Evolution*, eds. H.D. Regan, M. Worthing, Australian Theological Forum, Adelaide 2002, pp. 117–127.

Stoeger W.R., *Conceiving divine action in a dynamic universe*, [in:] *Scientific Perspectives on Divine Action: Twenty Years of Challenge and Progress*, eds. R.J. Russell, N. Murphy, W.R. Stoeger, S.J., Vatican Observatory Publications and the Center for Theology and the Natural Sciences, Vatican City and Berkeley, CA 2008, pp. 225–247.

Stoeger W.R., *Entropy, emergence, and the physical roots of natural evil*, [in:] *Physics and Cosmology: Scientific Perspectives on the Problems of Natural Evil*, Vol. 1, eds. N. Murphy, R.J. Russell, W.R. Stoeger, S.J., Vatican Observatory Publications and the Center for Theology and the Natural SciencesVatican City and Berkeley, CA 2009, pp. 93–108.

Stoeger W.R., *The big bang, quantum cosmology and* creatio ex nihilo, [in:] *Creation and the God of Abraham*, eds. D.B. Burrell, C. Cogliati, J.M. Soskice, W.R. Stoeger, Cambridge University Press, Cambridge 2010, pp. 152–175.

Stoeger W.R., *Emergence, directionality and finality in an evolutionary universe*, [in:] *Biological Evolution: Facts and Theories – a Critical*

Appraisal 150 Years after "The Origin of Species", eds. G. Auletta, M. LeClerc, R.A. Martinez, Gregorian and Biblical Press, Rome 2011, pp. 479–492.

Tracy T., *God and the causal structures of nature: Some puzzles*, this volume, 2013.

Van Till H.J., *The Fourth Day: What the Bible and Heavens Are Telling Us about Creation*, Eerdmanns, Grand Rapids, MI 1986.

Weinberg S., *Cosmology*, Oxford University Press, Oxford 2008.

Thomas Tracy
Bates College, Lewiston

God and the Causal Structures of Nature: Some Puzzles

Modern physics presents some intriguing new puzzles for understanding God's relation to the causal structures of nature. The way theologians think about the God-world relation is of course shaped by what we think we know about how the world is put together. Quantum theory provides a vivid instance of a physical theory that simultaneously deepens and disrupts our account of nature, confronting us with the limits of our ability to provide imaginatively satisfying models for fundamental physical structures. Talk about God has always been an exercise in the art of thinking about a subject matter that we know from the outset must exceed our comprehension. A striking feature of contemporary physics is that it too leads us to the boundaries of what we can conceive. Theologians may hear a familiar note in the title of the collected papers of the great theoretical physicist, J.S. Bell: *Speakable and Unspeakable in Quantum Mechanics*[1]. Perhaps there is some kinship between theology's ancient wrestling with the unutterable, and physics' modern encounter with the unspeakable, as both seek to advance understanding at the outer limits of what can be thought.

In this paper, I want to explore some theological questions that arise in this doubly difficult terrain. This is a foolhardy thing to attempt – I am a philosophical theologian, not a philosopher of physics,

[1] Cambridge University Press, Cambridge 1987.

and so I enter this territory with trepidation. My aim is to consider the impact of some concepts from contemporary physical theory on one of the major strategies in theology for thinking about God's relation to the natural order. The result, let me confess at the outset, is to raise questions that I am not sure how to answer.

1. Creation as God's Primary Act

In approaching the question about God's relation to the causal structures of nature, it is best to begin at the beginning. The doctrine of creation sets the fundamental terms of the God-world relation, and it has been developed in a number of different ways during it long theological history. I will focus here on the deeply rooted traditional view that God is the absolute ontological ground of all finite things, creating the world 'out of nothing'[2]. It may be tempting to read 'nothing' as a peculiar sort of 'something', a resistant negativity that God must overcome, and against which God forms a coherent world. But this would constrain the divine creativity, and the point of the *ex nihilo* doctrine is that God's creative act is limited by nothing outside God's own nature. The act of creation does not consist merely in making an ordered cosmos out of primordial chaos, and it does not depend upon any pre-existing unformed matter. Rather, God's creative act accounts for the very being of the creature, and apart from this act there would be nothing other than God.

This understanding of creation affirms that God does not require or need the creature; God is God quite without created things. The act of calling the world into being is a free and intentional divine initiative that God undertakes for the sake of the creature, because it is good for the creature to be. The existence of the universe is an ex-

---

[2] This discussion of *creatio ex nihilo,* occasionalism, and secondary causes draws upon material from my paper, *God and creatures acting: The idea of double agency*, [in:] *Creation and the God of Abraham*, eds. D. Burrell, J. Soskice, W. Stoeger, S.J., C. Coglioti, Cambridge University Press, Cambridge 2010, pp. 221–237.

pression of divine generosity, a sharing of the richness of God's being with finite things. This stands in contrast to accounts of creation as a necessary expression of the divine nature. One important historical rival to the idea of purposive creation, for example, is the Neoplatonic concept of emanation; the One does not choose to create finite beings, but instead gives rise to the world as an involuntary expression of its own overflowing plentitude of Being[3]. Classical doctrines of creation also differ from most forms of process theism, according to which God-and-world together constitute a metaphysically fundamental structure of creative becoming. Process theologies typically think of God *not* as the ontological ground or source of the world process, but rather as a uniquely comprehensive contributor to each new moment of its unfolding history[4].

The classical theological view, on the other hand, understands creation as an asymmetric relation in which God freely calls the world into existence and sustains it at every moment. God's creative act establishes a universal, immediate, and continuous relation of ontological dependence of creatures upon the creator. I started by saying that we should begin at the beginning, with creation. But it should now be clear that this concept of creation is not, in fact, tied to claims about the beginnings of things. It affirms, rather, that God is the world's ultimate sustaining ground and source. In this respect, a temporally eternal universe depends upon God just as profoundly as a universe that God called into existence at some moment in the past. Perhaps theoretical cosmology will lead us to the say that the early universe, regarded as a quantum entity, arose as a spontaneous fluctuation in a vacuum. But this highly dynamic primordial vacuum also depends for its existence upon God, and it should not be confused with the 'nothing' of *creation ex nihilo*. So when Stephen Hawking announces

[3] See Plotinus, *Enneads*, trans. S. MacKenna, Penguin, New York 1991.

[4] See A.N. Whitehead, *Process and Reality: An Essay in Cosmology*, eds. D. Ray Griffin, D.W. Sherburne, Free Press, New York 1978. For another version of 'neoclassical' theism see the writings of Ch. Hartshorne, e.g., *The Divine Relativity: A Social Conception of God*, Yale Univ. Press, New Haven 1948.

that quantum cosmology shows that there is no need for the action of a transcendent creator in order to get the world started, he simply misunderstands the theological concept of creation[5].

It is important to note that God's foundational creative action does not merely bring about particular *changes* in creatures, but rather brings about the *creatures themselves* along with all their properties and relations. The distinction between causing change and causing being is of fundamental theological importance because it marks an essential difference between the activity of God and of creatures. Created things operate as causes by modifying the properties of other existing things; this is true even of the activities we typically speak of as bringing something into existence (e.g., evolutionary processes creating new species; an artist creating a painting) or as causing something to cease to exist (by bringing about destruction and death). In these events, the configuration, or form, of some chunk of matter is 'trans-formed'. God's creative act, on the other hand, does not merely alter the properties of things, since apart from this divine action there is 'no-thing' to change. This divine action radically transcends the causality exercised by creatures; it is the metaphysically prior act of establishing and sustaining the existence of creatures. For this same reason, God's creative activity is radically immanent within the world: it is the immediate impartation of existence, an action that is closer to creatures than they are to themselves.

Creation in this sense is an action unique to God alone, and it is not something we can adequately describe or comprehend. As a shorthand we can speak of God 'causing' the existence of creatures. But the term 'cause' is used analogically here, and we are unable to spell out precisely what it denotes in this case. The concept of 'causing being' is a metaphysical black box; the most we can do is to indicate the effect of this divine action. Expressed as a counterfactual, if God did not do whatever it is that God does in causing the world to exist, there would be no world at all.

[5] S. Hawking, L. Mlodinow, *The Grand Design*, Random House, New York 2010, p. 180.

This understanding of creation has a crucial bearing on our central question about God's relation to natural causes. It follows from this view that God acts in every moment of the world's unfolding history. We cannot draw a distinction between events in which God acts, and others in which God does not; whatever occurs does so as a consequence of God's creative agency.

2. God as the Cause of Created Causes

At this point a difficulty arises. It might be thought that the doctrine of creation entails that God is the *only* agent at work in the world, and that the apparent operation of causes in nature is just the occasion for God's immediate act of specifying each new state of the universe. This view came to be known as 'occasionalism', and those who defend it typically contend that since God's creative action grounds the existence of every entity, along with all of its properties, God alone is causally efficacious. Imagine, for example, the mundane (but important) event of making a cup of tea. Given the doctrine of creation out of nothing, we might think that we should analyze this event in roughly the following way. At each moment God causes the flame, the teapot, and the water to exist and to possess each of their properties. Over a series of contiguous times, God successively instantiates a set of properties that vary in relation to each other (we might imagine a series of momentary actualities that add up to the entities familiar to us in our experience); the teapot and the water become warm, the proportion of water and water vapor changes, the vapor pressure increases, a teapot whistle occurs, and so on. Everything that happens here is done directly by God, and the causal roles played by created things are merely apparent, not actual. If God acts in consistent ways over time, then we will be able to formulate laws that describe our world, but these laws simply identify patterns of constant conjunction in God's creative activity from moment to moment. We have, then, the pleasing irony of a theologically motivated Humean view of causality.

This is not a purely fanciful metaphysical view. It has had a long history among Islamic thinkers (al-Ashari in the 10th century held roughly the view I just suggested), and it can be found in the philosophy of Malebranche. But Christian theologians have resisted the conclusion that their doctrine of creation entails that created things do not operate as causes in their own right. With Muslim occasionalists in mind, Thomas Aquinas remarked that 'There are some who have taken God's working in everything that acts to mean that no created power effects anything in the world, but that God alone does everything without intermediaries. For example, it would not be the fire giving heat, but God in the fire, and similarly in all other instances'[6]. He argues against this view, insisting that 'God's acting in creatures … must be understood in such a way that they themselves still exercise their own operations'. It is certainly within God's power to act as the occasionalists suggest, and enact by divine fiat all the changes that characterize the world's history; God could make the water boil quite without using the heat of the flame as a means. But Aquinas contends that God's generosity in creation includes granting causal powers to created things, so that in their interactions they affect and are affected by each other. As Aquinas put it, 'Divine providence works through intermediaries, … not from any impotence on his part, but from the abundance of his goodness imparting to creatures also the dignity of causing'[7].

This introduces the familiar scheme of primary and secondary causation, and marks a distinction between what God does directly and what God does indirectly through the activity of created things. It is helpful to link this distinction to the metaphysical difference we just noted between divine and created agency. Creatures, we said, operate as causes solely by inducing changes in other things. God, on the other hand, acts more radically to 'cause' the very existence of creatures. This primary creative act is always direct; it is the immediate

[6] *Summa Theologica*, I, 105,5.
[7] *Ibid.*, I, 22,3.

giving of being. But God also causes change, and may do so either directly or indirectly. If God causes the pot of water to boil spontaneously, then God acts *directly* to bring about a change in the properties of an object. In this instance, the effect of God's direct action departs from the ordinary course of nature, and so is a miracle in the Humean sense[8]. Ordinarily, however, God brings about particular changes in the world indirectly, acting through creatures as causal intermediaries (e.g., God heats the water by means of the flame). By establishing and sustaining an order of natural causes, God produces the unfolding series of events that constitutes the natural history of our world.

It follows from this that specific events, to the extent they follow from the causal structures and initial conditions that God has put in place, can be attributed to God as indirect divine acts. For example, the evolution of rational moral agents, and perhaps the existence of each of us here and now (depending upon the details of the evolutionary story and its theological interpretation), can properly be described as an intentional action of God. This conclusion is not in the least undercut if we acknowledge that this action was carried out through a staggeringly vast series of natural causes developing over billions of years, rather than by a specific divine intervention in biological processes. One consequence of this view is that, *contra* intelligent design theory, we need not insist on the incompleteness of biological explanations in order to affirm that God's purposes are at work in the history of life.

The scheme of primary and secondary causation, then, provides a powerful conceptual structure for understanding God's action in and through the natural order, and it continues to be widely invoked in theology. I want to consider what becomes of this conceptuality when we bring it into contact with new ways of thinking about nature that emerge in contemporary physics, especially in quantum me-

[8] We will see in a moment that, in an indeterministic universe, God's direct acts in the world need not take the form of miraculous interventions in otherwise complete natural causal series. Further, the language of 'intervention' is problematic for a number of reasons, not the least of which is that, on the account of creation I have given, God is always engaged in every event, so that there are no events that are simply 'natural'.

chanics. In order to locate the key issues, it will be helpful to say a bit more about the metaphysical underpinnings of the idea of secondary causation.

3. Causal Powers

The theological understanding of created causes that we have been considering is tied to an ontology of substances and their properties. Substances are bearers of properties (the substance 'stands under' them – *substantia*), and properties may be either essential or accidental to a substance. Roughly speaking, a property is essential to an individual substance if that individual could not exist without that property; a property is accidental if the individual could exist either with or without it[9]. For example, it is typically said that at least some categories of kind identity are essential to the individuals who instantiate them. I could not exist except as a member of the human species; e.g., I could not have been a bat, though there are any number of other properties that I do in fact possess but could have lacked (e.g., the property of having lived in an old house with bats in the attic).

Kinds are marked out by a common set of essential properties, including various active and passive causal powers, that is, capacities to bring about changes in other substances and to be changed by them. These causal powers are systematically linked in the ordered network of properties that the substance possesses as an instance of its kind. Using a modern example, carbon dioxide traps heat in the atmosphere because its molecular structure absorbs and re-emits electromagnetic radiation in the infrared. On this view, causal relations and causal laws are rooted in intrinsic properties of the things that God creates. By virtue of what the entity *is* it has a disposition to display characteristic behaviors under specific conditions, and this causal disposi-

[9] The modal concept invoked here not simply logical necessity rooted in conceptual relationships *de dicto*, but rather metaphysical necessity, a necessity *de re*.

tion can (at least in part) be described by set of counterfactual generalizations about how the entity would act or react. Laws of nature are descriptive of the essential causal capacities of things; they are prescriptive only insofar as they articulate the logos of creation as conceived in the mind of God[10].

4. Physics and Metaphysics

This substance ontology was based on metaphysical interpretation of familiar objects and events at the macroscopic level. The quantum realm, however, resists characterization in these terms, and there is no agreement on what new account we should give of the fundamental structures in the world that give rise to the behavior described by the quantum formalism. This is the first point to make about quantum mechanics: the theory can be interpreted in a striking variety of different ways, and there is no unproblematic path from quantum physics to metaphysics.

This interpretive pluralism is due at least in part to the irreducibly statistical character of the quantum formalism. According to the uncertainty principle, it is not possible to specify with complete precision all of the properties of a quantum system. For example, electrons have a determinate mass, charge, and spin magnitude. But we cannot simultaneously know the electron's position and momentum or its spin orientation on more than one axis. Until we make a measurement on the system, we must describe its state as a probabilistic mix of incompatible values for these properties. The distribution of likely outcomes on measurement is well-defined, and it varies in conformity with the wave equation, which develops smoothly over time.

---

[10] For some of the extensive and diverse literature on causal powers and natural necessity, see: N. Cartwright, *Nature's Capacities and Their Measurement*, Oxford University Press, Oxford 1989; R. Harre, E. Madden, *Causal Powers: A Theory of Natural Necessity*, Basil Blackwell, Oxford 1975; B. Ellis, *Scientific Essentialism*, Cambridge University Press, Cambridge 2007.

Each successive probabilistic state of the system determines the next according to the Schroedinger equation, but the states described by this equation are superpositions of alternative possibilities. When we interact with the system to conduct a measurement, however, a specific value is obtained for the property we have chosen to observe (e.g., we might run the electron through a magnetic field that records its spin orientation). It is at this point, in the transition from a superposition of possibilities to a determinate outcome, that the causal story told by quantum mechanics is incomplete. The theory, at the level of the quantum formalism, does not provide causally sufficient conditions for the selection of one of the probabilistically structured potentials rather than another.

This puzzling but powerful theory challenges our ability to form an intelligible picture of the world, and interpreters have generated a wide variety of different suggestions regarding what it reveals about the deep structures of nature. Discussion of interpretive questions has been marked by sharp disagreement from the earliest days, e.g., in the extended debate between Bohr and Einstein about the completeness of quantum mechanics. It is important to acknowledge that both deterministic and indeterministic interpretations can be given, though Bell's theorem and EPR-type experiments have shown that we cannot reproduce classical local determinism at this level of nature. For example, David Bohm's ingenious reconstruction of quantum theory provides particles with a complete set of determinate properties, position being basic, but introduces guidance equations, or 'pilot waves', that maintain instantaneous correlations at space-like distances[11].

If we want to think about the possible theological implications of quantum mechanics, we cannot avoid tying the discussion to one or another of the available interpretative approaches, and each will present somewhat different issues about God's relation to quantum

---

[11] D. Bohm, *A suggested interpretation of quantum theory in terms of hidden variables, I & II*, "Physical Review" 1952, no. 85, pp. 166–193. Also see, *Wholeness and the Implicate Order*, Routledge & Kegan Paul, London 1980.

processes. I want to reflect briefly on indeterministic interpretations, which seem to predominate in current discussion. There are multiple candidates here, including several flavors of Copenhagen interpretation. In response (in part) to the measurement problem, there also are a number of more recent approaches: e.g., decoherence, GRW (Ghirardi, Rimini, and Weber), and spontaneous collapse theories. These views differ significantly from each other, but for my broad purposes it is enough to note what they share: 1) they accept the completeness of quantum mechanics, and so deny that every property of quantum systems has a determinate value at every moment, and 2) they accept the under-determination of the transition from the superposition state to a determinate outcome.

These features of quantum systems (i.e., indeterminateness and underdetermination), along with their nonlocal character, disrupt the classical theological scheme of divine action through secondary causes. Let me comment briefly on each of these issues.

5. Indeterminateness

If we suppose that there are no hidden variables which would allow us to specify a value for each property of a quantum system at every moment of its evolution, then uncertainty about these values is not merely epistemic, but rooted somehow in the quantum entity itself. An electron just *is* an odd mix of determinate and indeterminate properties, e.g., it has a determinate mass distributed probabilistically across a region of space; it has a determinate spin magnitude in a superposition of incompatible orientations. It hardly needs to be said that this is not the sort of entity envisioned by classical theories of secondary causation, and it appears that the metaphysics of substances with essential intrinsic properties linked in a set of causal dispositions will need to be fundamentally revised, if it can be sustained at all.

Bruce Gordon makes a case of this kind against 'naturalism' as a whole, which he defines as the view that 'the sum and substance of everything that exists is exhausted by physical objects and processes and whatever is causally dependent upon them'[12]. We need not reproduce his full argument here; our main interest is in his contention that necessitarian theories of causal law, such as causal power and counterfactual views, break down at the quantum level. Let me just sketch two interconnected difficulties that he identifies.

First, a necessary condition for being a physical object is that an entity must possess well-defined properties that allow it to be uniquely identified, or individuated. Typically this is done is through spatio-temporal location. If quantum mechanics is complete, however, then the fundamental constituents of physical reality cannot be picked out as discrete objects. Their intrinsic properties (like the mass and charge of an electron) are identical with all other particles of the same type, and therefore do not identify unique individuals. Their quantum state dependent properties (like position and momentum) are indeterminate, and therefore also do not uniquely identify a single individual. So quantum entities fail to meet one of the necessary conditions for physical objecthood[13].

Further, causal power theories understand causal roles to be a function of the law-like relations of the essential properties of objects. These properties are expressed as dispositions to produce characteristic outcome states under appropriate stimulus conditions, and they thereby ground causal laws. Once again, however, if quantum theory is complete, then some of the properties of quantum entities are indeterminate prior to measurement, and therefore are not amenable to a causal powers analysis[14].

We might immediately reply to this second point by incorporating the stochastic properties of quantum entities into a revised theory of causal powers. Rather than requiring determinate properties linked

---

[12] *A quantum-theoretic argument against naturalism*, [in:] *The Nature of Nature: Examining the Role of Naturalism in Science*, eds. B.L. Gordon, W.A. Dembski, ISI Books, 2011, p. 181.

[13] *Ibid.*, p. 182.

[14] *Ibid.*, pp. 181, 203.

by necessitating relationships expressible in single-outcome counterfactual conditionals, this new view would give a crucial role to non-necessitating, probabilistic causality. The quantum entity would then be understood as an evolving structure of potentiality. Its essential nature is defined not only by its intrinsic determinate properties, but also by a set of well-defined probabilistic propensities, and its causal powers include its capacity to generate with precise likelihoods just these outcome states. The causal role of the quantum entity would be spelled out in a set of counterfactual conditionals which, for some of its properties, specify not what *would* happen, but rather what *might* happen, articulated stochastically.

6. Nonlocality

The flaw in this otherwise promising strategy is that our revised story about causality once again runs up against the problem that quantum entities cannot be adequately individuated even as discrete bearers of stochastic properties. The electrons in a two electron system (like a helium atom) are qualitatively identical, and they must be described by a single wavefunction. But according to the Pauli exclusion principle, this wavefunction must be anti-symmetric, so that they cannot have the same quantum number. Their joint state cannot be represented as the sum of objective probabilities possessed by each electron considered by itself; rather, they must be described in relation to each other. They become differentiated as distinct individuals only when one of their probabilistic properties collapses to a determinate value, simultaneously fixing the value of its anti-correlated counterpart. This connection persists even when they are separated at space-like distances. In EPR-type experiments, physicists make separate measurements on two particles. But the relationship between the particles cannot be captured by the idea that they are localized bearers of causal powers. Rather than there being a transmission of causal influence between discrete individuals, these particles constitute a relational system, a whole that somehow maintains instantaneous correlations of its separated components. There appears to be no intelligible causal analysis of this situation, and yet current physical theory suggests that it is ubiquitous in nature.

These considerations make it difficult to be a realist, even a chastened and critical one, about quantum theory, and they suggest that we might think of the theory simply as a useful construct that guides our actions and expectations. On a view of this sort, quantum mechanics would provide a reliable mapping of predictive regularities, but it does so without supplying an account of what gives rise to them. The theory would function as a probabilistic calculator, telling us what is likely to happen under specified conditions. But the object language of the theory (its talk about the various inhabitants of the 'particle zoo') can be empirically adequate, in the sense that it accounts for all current observations, even if we are agnostic about whether this language correctly describes real items in the world, i.e., about whether it is literally true[15]. This is not to deny that there *is* something we interact with in experimental physics; it is just to say that the theories we construct about this level of structure in nature may turn out not to tell a descriptively, or representationally, helpful story about what that 'something' is. The language of particles and properties has its value not as a description of the quantum stuff we encounter in the lab (and elsewhere of course), but as a conceptual model that can accommodate current observational data, map relationships, predict results, shape ongoing inquiry, and so on.

If we were to embrace this cautious anti-realism, then we would have to abandon the theory of causal powers. Alternative accounts of causation are waiting in the wings, however. Regularist views hold that causal statements and causal laws just identify recurring sequences in experience. On this view, secondary causes with active

[15] On the idea of empirical adequacy see B. van Fraassen, e.g., *To save the phenomena*, "Journal of Philosophy" Oct 1976, 73, no. 18, where he argues that 'the language of science should be literally construed, but its theories need not be true to be good' (623).

powers of their own disappear into predictable patterns of observable events. This metaphysical modesty is appealing, but I want to post a warning that if we import regularism into theology, it might bring with it the second coming of occasionalism. This difficulty comes into view when we turn to another of the striking features of quantum mechanics that I initially noted.

7. Underdetermination

The interpretive approach that we are considering here understands the transition from the superposition state to a specific outcome (or to localization of the wave function, as the case may be) as causally discontinuous. Although the wave equation evolves smoothly and deterministically over time, its collapse to a determinate value for the measured property is not fully determined by anything in the prior history of the system. The outcome will have necessary but not sufficient conditions in antecedent events, and as a result, the causal structure of the universe (whatever it might amount to) will not be closed and complete.

At this point a puzzling theological question arises: what is God's relation these underdetermined quantum transitions? The first point to make is that God acts as their creator, just as with every entity and event in the history of the universe. The puzzle in this case, however, is to conceive of *what* God creates. In presenting the idea of creation, we made a distinction between the divine action of giving being, which is always direct, or immediate, and the act of causing change, which for God may be either direct or indirect. The reply to occasionalism called upon this distinction to contend that God may directly empower the existence of a creature along with all of its properties without also directly causing it to acquire those properties – that may be left to other creatures operating as causes in their own right. Now, however, we find ourselves considering a class of exceedingly odd entities, these bundles of evolving potentiality, which undergo transitions that do not have sufficient secondary causal conditions. In this context, can we any longer make a distinction between what God does directly, in giving being, and what God does indirectly, in acting through creatures to cause change? It appears that the outcome state of a quantum transition must either be determined by God or by nothing at all.

The second of these two options is particularly intriguing and puzzling – it is the idea, roughly speaking, that God leaves quantum transitions 'up to chance'. God in this case would have the temerity to defy Einstein, and massively play dice with the universe. It is important not to be misled by the familiar expression 'leaving it up to chance', as though chance were an odd sort of cause, a process or mechanism that can be relied upon to produce an outcome. This is characteristic of the way chance works for *us*. We roll the dice, unable to predict the result because we cannot calculate which outcome will follow from the initial conditions and the relevant causal laws. But this is just epistemic chance, reflecting the limits of our cognitive powers; the underlying physical processes may be perfectly deterministic. We also can make use of true under-determination in nature, if there is any, by arranging devices that amplify chance transitions to produce observable results – this happens in particle physics labs all the time, and it also occurs (quite without human design) in the natural world, e.g., in some kinds of genetic mutation. But these uses of chance are not available to God, since God does not have the relevant cognitive limitations and does not encounter chance as a given property of the world.

So how could the creator leave some aspects of creation 'up to chance'? The puzzle here is that God would bring a creature into existence without fully specifying what that creature shall be. It is safe to say that it makes no sense to assert that God creates something that is nothing in particular. But perhaps we could think of God creating units of structured potentiality – or in light of non-locality, webs of interwoven potentialities – without specifying which of their future states is actualized. The created world would then spontaneously undergo transitions to new states, permitted by the stochastic structure of the potentialities God called into being. Peter van Inwagen appears to propose something like this in his discussion of 'The Place of Chance in a World Sustained by God'[16]. He contends that in an indeterministic universe some states of affairs might neither be determined by secondary causes nor by God[17]. Indeed, he suggests that God might choose not to determine even the initial state of the universe. Perhaps God could issue a disjunctive creative decree – 'Let either X or Y be' – and just leave it at that. 'It does not seem to me to be logically or metaphysically impossible that God should decree that either X or Y should be without decreeing that X should be and without decreeing that Y should be'[18]. Suppose that the result of this divine decree is that X springs into existence. It is tempting to ask, 'If God does not cause X to exist rather than Y, then what does?' There can be no answer to this question, and its persistence reflects just how deeply our thinking is informed by the principle of sufficient reason. Defenders of the cosmological argument, for example, would find their explanatory aspirations blocked at the crucial point: the appeal to a self-existent creator of the world would not after all provide a sufficient reason for the existence of *this* universe.

The alternative is to say (as a first approximation) that God specifies the outcome of naturally underdetermined transitions in quantum systems. This has the advantage of preserving the principle of sufficient reason; now every change has an explanation either in secondary causes, in finite intentional agents, or in God. This mode of divine action goes beyond sustaining the existence of creatures, on the one hand, and acting indirectly through them to produce their various effects, on the other. Here God acts directly to cause particular changes in the world, bringing about events that would not (or might not) have occurred had God not so acted. It is worth noting in passing that this

---

[16] In *Divine and Human Action*, ed. Th.V. Morris, Cornell University Press, Ithaca, NY 1988.

[17] *Ibid.*, p. 224.

[18] *Ibid.*, p. 227.

provides a means by which God can profoundly shape the course of the world's history without in any way disrupting whatever natural causal structures are in place; it therefore constitutes a form of non-miraculous special, or particular, divine action. It also, however, entails that God is massively and pervasively engaged in the direct production of events in the world, determining all those states of affairs that the natural order leaves determinable (and that are not determined by finite intentional agents, if there are any who are able to act in this way). This leads to a number of theological worries, including questions about determinism, moral freedom, and the problem of evil. Determinism and the problem of evil, you will be relieved to hear, must wait for another time. But since I began this discussion by discussing occasionalism, I want to finish with this topic.

8. Occasionalism Redux?

Does this view of God's action in underdetermined events lead us back to occasionalism? In order to see why one might think so, let me briefly pull together the threads of the discussion so far. Quantum mechanics, we have seen, requires that we rethink traditional accounts of secondary causes as substances possessing various intrinsic causal powers associated with being the kinds of things they are. Quantum non-locality in particular cannot be understood in terms of classical causal concepts, and leaves us with stochastically lawful, but unexplained, correlations. Suppose that in the face of this we adopt a regularist interpretation of these causal laws. The object language of quantum theory is then understood as a conceptual device that aims only at empirical adequacy, not at true description of objects in the world. As a result, we cannot now say that God produces these effects by means of the operation of quantum entities as secondary causes, because we no longer have an intelligible concept of the bearer of such causal roles and powers. If, in light of the under-determination of quantum transitions, we now assert that God directly brings about each outcome state, then quantum phenomena are an immediate expression of God's activity–and so is the world of macroscopic objects, properties, and relations that is built up on this base.

This certainly sounds like the triumphant return occasionalism[19]. On this account, our familiar world consists of a vast aggregation of instantaneous property realizations immediately produced by God. God is the only cause of change as well as of being, and natural laws simply identify recurring patterns in this divine action. The world *appears* to contain finite objects bearing intrinsic causal powers, but this is merely an appearance. As Aquinas recognized, in an occasionalist world created things do not 'exercise their own operations'. They lack, we might say, a structure of active being that is their own, and in this respect the occasionalist's God stops short of positing the creature as an other, differentiated from God's own activity.

Occasionalism has its advantages, and I have argued here only that it is not an inevitable consequence of classical conceptions of creation. If we think it important to resist occasionalism, however, then there appear to be at least two strategies for finding a way out of these puzzles about secondary causation. One is to come up with a revised causal powers view at the quantum level, perhaps along the lines that I suggested briefly above in speculating about structures or networks of non-local created potentiality. The second is to argue that secondary causation is an emergent phenomenon. But note that this would involve a unique sort of emergence. In this case we would not be saying that secondary causal processes at a lower level in nature provide the base for new kinds of supervening causal processes at a higher level. Rather, we are imagining that direct divine action without created causes at the quantum level would provide the base for the emergence of secondary causes at higher levels. That is, there would be no entities with causal powers underlying quantum phenomena, but God's activity at this level would give rise to such entities at higher

[19] And just this is the conclusion that Bruce Gordon, in an unpublished paper, draws from his critique of naturalism.

levels. This would involve a transition from occasionalism at the micro-level to the emergence of intrinsic causal powers at higher levels. This is an intriguing idea, but it remains shrouded in one of the central puzzles of contemporary physics, namely, how the classical world arises from quantum processes.

I said at the outset that I was setting out in this paper to ask some questions that I cannot readily answer, and it is safe to say that I have succeeded in that. Here is a modest conclusion: we theologians may wish to keep using the primary/secondary causation scheme, but if we do so, we ought to recognize and grapple with the fact that its metaphysical foundations have been undermined, and it is not clear how to provide a new basis for a comparable way of relating divine and created causes. We can take some comfort, however, in the knowledge that theology is not alone in finding that old ways of thinking about causality have come unstuck in the encounter with quantum mechanics.

Willem B. Drees
Leiden University

God as Ground? Cosmology and Non-Causal Conceptions of the Divine

Some theistic arguments envisage God as the 'First Cause', as a being temporally pre-existing relative to the cosmos, of all created reality. For the natural order, the universe, God's activity would be the answer to the question 'Why is there something rather than nothing?' Such a 'first cause' view of God may be integrated with cosmic natural history, with a 'Big Bang' as its initial event. However, at the boundaries of physical cosmology our notions of time and causality seem to break down. It therefore is disputable whether a 'Big Bang' as the limiting event of standard cosmology provides a stable model for 'the first event' in future theories. Nonetheless, even when theories probe further and thereby change key concepts of space and time, the mystery of existence remains. Perhaps, the transcendent – if there is such a thing – is not to be thought of in temporal terms. Given the role of mathematics, and its independence from the physical dimensions of time and space, one might try to draw on mathematics and logic to imagine 'transcendence'. Axioms are not so much the cause of the theorems, but rather the formal ground of all subsequent theorems.

What would it mean to speak of God as ground? Do 'God as ground' theologies offer a possible way beyond the conceptual challenges at the limits of cosmology? Would it be a proper alternative for the ontological dualism of classical theism and the indiscriminate valuation of reality that seems to be a consequence of pantheism? And if one goes down this route, what might be lost? It has been commented on Anselm's ontological argument (which also argues for God along the lines of logic rather than of causality) that the argument in its modal form can be considered to be a proof of the non-existence of God, as the type of existence ascribed to the being 'greater than which nothing can be conceived' is similar to that of mathematical objects – meaningful but without reference.

In the following contribution we will begin (1) with the tradition of natural theology, within which 'design' and 'first cause' arguments have their place. We then turn to (2) the Big Bang theory and its limitations. The development of theories (3) 'beyond the Big Bang theory' gives us a context to reflect further upon the idea that there is a *first* cause. This provides the context for reflecting upon (4) a different way of conceptualizing God, as 'ground' rather than as 'cause'. The final section of this contribution is (5) a reflection on the nature of theology as a particular type of human construction.

1. Natural theology

One prominent style of the European engagement with the natural sciences in the seventeenth to nineteenth centuries has been 'natural theology'. Connecting theological ideas with insights from the natural sciences has been especially widespread in the United Kingdom. John Ray's *The Wisdom of God, manifested in the works of creation* (1691) was full of observations in natural history, and influenced Carl Linnée (Linneaus) who designed the major classification of biological species. William Paley's *Natural Theology, or Evidences of the Existence and Attributes of the Deity* (1802) piled up examples of the intricacy and purposefulness in organisms as evidences of the intelligence and goodness of the Creator. Such natural theology was not just an apologetic argument for religion. Showing how science fitted a theological understanding was as much apologetics for science, which was not yet as useful and respected as it became later (Brooke and Cantor 1998, 148-161; Harrison 2008).

In the domain of biology, religious explanations of functionality in organisms became superfluous with the rise of evolutionary explanations, especially since Charles Darwin's *The Origin of Species* (1859). Some Christians were so strongly attached to the argument from design that they opposed evolutionary explanations – a strand that has had its own evolution in the subsequent century and a half (Numbers 1992). Many other Christians have accepted evolutionary explanations.

Some of those Christians who accepted Darwinian evolutionary explanations, used the design argument at a more general level. One might envisage that God has designed the conditions and laws of nature such that the evolutionary processes could bring forth organs and organisms adapted to many different circumstances. Thus, theologians and lay persons might speak with awe and reverence of evolved nature, and of the Creator who made it all possible. This higher order design tradition, accepting explanations based on the laws of nature, marveling at their remarkable fertility, has continued in discussions on modern cosmology, especially in a patterns of reasoning called 'anthropic arguments'. Though this is not a return to the geocentric understanding which had humans close to a spatial center, advocates of such reasoning suggest that the universe seems designed for the purpose of bringing forth humans (or, at least, intelligent living and loving beings).

The argument may seem unlikely, as the Universe is enormous in size when compared to human dimensions. The age of the Universe is a million times the age of human civilization. However, other things being equal, the vast age and size of the Universe might be related to our existence. We need carbon and oxygen. As heavier elements are formed in the interior of stars, we needed several generations of stars before we get an environment that has the right chemicals. Evolution took another couple of billions of years to produce complex, intelligent, observing, and amiable beings – us.

Turning this description upside down, it is argued that intelligent observation by natural beings is only possible after a couple of billion years. Thus, biological beings can only observe a universe that is something like ten billion years old. Along these lines one might invoke a weak anthropic principle (WAP) to 'explain' more or less the observed age of the Universe, given our existence. The same reasoning applies to the initial conditions assumed in the Big Bang model. Our existence depends on properties of the universe, as if it were all made so that we could arise.

In my opinion, anthropic principles do not function properly in scientific explanations (Drees 1990, 81-89). Either the contribution is trivial, as is the case for the so-called Weak Anthropic Principle, or the contribution is metaphysical, as is the case for various strong anthropic principles. The arguments do not work from science to metaphysical conclusions, such as the existence of a human-loving God. Rather, the anthropic principles *presuppose* certain metaphysical positions which, once accepted, may carry with them certain views of the Universe.

So far for a very brief history of 'the argument from design', first in the context of biology, later in the context of cosmology. This approach was typical of 'natural theology', a strand in the Christian tradition. This argument is philosophically contested (e.g., Manson 2003). And it is also religiously not appreciated by all. Not all theists are looking for arguments that connect faith and cosmology so intimately; a topic we will return to in the final section of this paper.

There is a different class of arguments, called 'cosmological arguments', which are less dependent upon specific characteristics of organisms or the universe, and hence seems less vulnerable to progress made in science. A classic exponent of such arguments has been Thomas Aquinas (13th century), who in his *Summa Theologiae* (1a 2, 3) presented 'five ways' which would be arguments for the existence of God. The first one is based on movement (or change): we see movements, and we see that movement is generated by other movement. If we follow the causal chain back in time, we either have an infinite chain or a 'first mover', which is itself not moved by anything else. If we don't accept that that there has been an infinity of times and states before the present, we thus come to a first, unmoved mover. In modern terms, this might be rephrased in terms of energy and momentum, conserved quantities which don't appear out of nothing.

Chains of causes or explanations have a certain natural appeal, reflected in questions such as 'Where does it all come from?' and 'Why is there something rather than nothing?' We can imagine that the world would have been different or would not have been at all – reality is contingent. In arguments such as those of Aquinas, this radical contingency is resolved by postulating a necessary being: God as *causa sui*, not dependent upon anything else, a first cause; God as unmoved mover. Critics of the argument find this a problematic move – all the time one insists that explanations are needed. These push one to further and further abstractions – and suddenly, one argues that no further explanation is needed, as something is postulated as necessary, its own explanation. Religious critics have another concern: that 'first mover' may have been long ago; how to avoid postulating a God who started the sequence long ago and was irrelevant thereafter.

Design arguments are about specific properties of organisms or of the universe, and thus nourished by observations about nature. They may thus have some affinity with science, but are also challenged by scientific explanations of those specific properties. Cosmological arguments such as the argument that the sequence needs a 'first cause', are less dependent upon science. Nonetheless, the 'first cause' argument has its interactions with modern cosmology, as we will explore in this paper.

Science plays less a role in this argument, compared to the design argument. For the argument about God as 'the first cause' certain broad conceptions of explanatory or causal sequences are needed, as well as the 'observation' that there is something rather than nothing, but no specifics of organisms or of the universe, or at least, so it seems.

2. The Big Bang as First Cause?

The Big Bang Model and its Limitations

The Big Bang model of the Universe combines observational data such as those on the expansion of the universe, and theories such as general relativity as the fundamental theory of space, time and gravity, and the best available theories on matter (fields and particles). Others in this volume will provide a far more detailed discussion. For the present purpose it seems sufficient to summarize the theory and its limitations in very broad strokes.

The theory suggests that the universe has been expanding, as shown by the movement of galaxies relative to each other. Counting backwards, this suggests that the observable universe has had a very dense and hot state. Going all the way back, there seems to have been a moment of infinite density and temperature, a beginning of the universe. Current evidence suggests that this moment has been almost 14 thousand million years ago.

Conceptually, we have to distinguish between two aspects of the Big Bang theory. (a) The Big Bang theory is a theory about the *development* of the universe during billions of years. This theory has been very successful, corroborated by increasingly precise observations on the distribution of various types of atoms in the universe, the cosmic background radiation, and much more. (b) In line with the name, the Big Bang theory is perceived as a scientific theory about 'the Big Bang', the initial Singularity. However, this is a mistaken view of the Big Bang theory: the theory does not reach that far. The theory is not about the Big Bang but about the subsequent evolution of the universe.

The Big Bang as a singularity (event of infinite density and temperature) lies outside of the domain where the theory can be trusted, for two reasons. (i) The theory uses our knowledge to look back in time. This study of our past is very successful. As Steven Weinberg in his popular account *The First Three Minutes* (1977) made clear, cosmology and particle physics became intertwined, as the cosmological consequences of advanced theories in particle physics became a major testing ground for particle theories. A theory that describes the observable universe from the present way back until well into 'the first three minutes' is incredibly impressive. For most of the time, from the present back, scientists draw upon physics that is well tested. But somewhere deep within the first second, we don't know how matter behaves: the particle physics for such high energies is not tested within human laboratories, and the theories are speculative, such as those about 'superstrings'. As we aren't sure how matter behaves, extrapolation towards earlier times becomes equally speculative.

(ii) Closer to the Singularity comes a moment, presumably the 'Planck Time', a number constructed from fundamental constants of quantum theory and gravity, about 10^{-43} seconds after the initial Singularity, where the combination of these theories breaks down. General Relativity Theory must be replaced by a quantum theory of gravity. Some current ideas will be touched upon below. As General Relativity Theory is our best theory about space, time and gravity, the uncertainity aboput the theory implies that we aren't sure anymore how to think of 'time', and whether the concept is still meaningful. And once 'time' is no longer meaningful, it becomes unclear what can be ment by 'before the Planck Time' – and hence by the Singularity that is supposed to be the earliest moment of time.

If one were to continue backwards in time, the initial Singularity itself would be a third limit, where General Relativity, the theory about spacetime, breaks down. However, as this limit lies beyond the Planck Time, and thus in a realm where general relativity has to be abandoned anyhow, it is not clear in what sense this limit might be relevant at all. This cannot be decided without considering the actual theories of quantum gravity that have been proposed. Whereas the first and second limit we encounter when going back in time are limits to our present knowledge, the third seems to be an edge, an ontological discontinuity – but it is hidden behind the other two.

A Religious Interpretation of the Big Bang?

Whereas the Big Bang model treats later states of the universe as developing out of the earlier ones, the limiting 'event' at 't = 0' has no predecessors within this model. Would this be 'the moment of creation'? An early claim of this kind was made by Pope Pius XII, in a speech to the Pontifical Academy of Sciences on 22 November 1951, appropriating the Big Bang in the context of the classic cosmological argument. George Lemaître, the Belgian astronomer and priest who was one of the original proponents of the model, was very unhappy about the way Pius XII used the Big Bang theory as physical proof of creation (McMullin 1981, 53).

Others, such as the astronomer Fred Hoyle, disliked the Big Bang theory, precisely because it seemed to suggest such an initial moment and thereby opened the door for a theistic understanding of the universe. Again others have argued that the Big Bang theory is religiously neutral. One might argue, for instance, that the apparent 'initial moment' can be placed an infinite time ago if one redefines the parameter 'time'. Or that the initial singularity is a conceptual limit of the model, rather than a description of an actual event. Others suggest that this is not what religion is about, for instance by distinguishing 'how' and 'why' questions, causal and intentional explanations, or perhaps even facts and values or symbols.

An interesting modification of the 'neutrality' position has been formulated by the philosopher Ernan McMullin. He does not believe in 'support' from the Christian doctrine of creation for the Big Bang model, or the other way round. But the Christian '*must* strive to make his theology and his [scientific] cosmology consonant in the contributions they make to this world-view' (McMullin 1981, 52). This consonance is 'in constant slight shift'. We will come back to this quest for consonance in the concluding section of this contribution.

A Beginning in Time or a Beginning of Time?

If the Big Bang were taken as the initial moment of existence, philosophical questions arise about the understanding of time. Basically, we seem to face two alternatives – to treat it as a beginning of the universe *in* time, or as the beginning *of* time.

The first alternative is like any beginning. Somewhere on the timeline my life has begun, this piece of art was made, this city came into existence. It may be hard to define a precise moment for a city, or even for a human being, but the basic idea is that beginnings can be located on a pre-existent continuum. And hence, for any beginning one may ask what went on before: my parents met, and longer before, my grandparents, or the artist had a certain idea, or there was a trade post, and so on. If that is how one tries to envisage the beginning of the universe *in* time, the suggestion of a 'before' arises. A *regressum ad infinitum* may arise; contingency is relocated to earlier and earlier stages in time.

Some cosmological models have tried to embed the Big Bang model in a larger picture of successive phases of the universe, whether cyclical (each phase collapsing into a dense state like the beginning) or as a two phase process, contracting since past infinity and expanding for all future times. By treating time as a given background, such models go against the mood of general relativity theory.

Another variant is to assume that there was no universe before 't = 0'; the universe began in time. Time preceded the universe, but for an infinite period the show had not started yet. This too results in a similar objection, that the flow of time becomes a given rather than a feature of the universe. It also makes it hard to imagine any reason as to why the universe did start at that particular moment in time, rather than any earlier one, and hence, given the infinity of earlier times, why it had not started again an infinitely long time before.

A similar problem was raised by Augustine in Book XI of his *Confessiones*, as he contemplated the creation. What was God doing all those ages before God created the world? Augustine first says that he is not making the joke that God was making hell for those who ask such questions. His serious answer is that the question is wrongly posed; time is bound up with movement and change, and hence with the created order. When there was no creation, there was no before, and hence no reason to ask what God was doing before He created the world. Time came into being *with* the created order (*creatio cum tempore*, rather than *creatio in tempore*).

Creatio cum tempore expresses a potentially powerful way to re-envisage 't = 0', by avoiding the idea that time is a container, from past infinity to future infinity. Perhaps we have to think of the beginning of the universe as the beginning *of* time. And, so some theoreticians hope, perhaps the correct quantum theory of gravity implies a quantum cosmology that does this job well. In that sense, the scientific quest for understanding fundamental physics is intertwined with the one for understanding the very beginning of our universe. If one accepts this idea of a beginning *with* time the contingency of a beginning at some apparently arbitrary moment of time is avoided. Of course, it may well be that some features of the universe that arises are still contingent rather than explained by the theory – and the theory itself may not be the only possible one. Thus, forms of explanatory (natural or logical) contingency may remain.

3. Beyond the Big Bang Theory

A Plurality of Approaches

The Big Bang theory is a very successful scientific theory about the evolution of the universe, but it does not explain or describe the Big Bang itself, but deals with the evolution of the universe thereafter, and especially after the Planck time. We seem to need new physics in order to push the explanatory quest in cosmology further. Success in this explanatory quest seems to be the main ground where such new physics might prove its potential.

Though all science is to some extent a human construct, certain results are so well corroborated and used in so many different ways (e.g. the Periodic Table of atomic elements) that the constructed nature of such knowledge does not diminish its claim to truth, understood realistically, at least for the domain or scale of resolution where atoms are an adequate model. However, uncertainty creeps in when one moves on to finer scales and start speaking of quarks and gluons, and even further down, of superstrings. This uncertainty results in a plurality of research programs in 'speculative cosmology'. Observations and theories at later times or accessible scales are not necessarily consistent with just one model of the underlying reality. Within cosmology there may be genuine underdetermination of theories (or, as they often are, outlines of theories) by the data. For instance, when we consider the issue of the beginning of time, the approach formulated by Hartle and Hawking is quite different from the ones chosen by Vilenkin and by Penrose, and again different from the ones by Linde and by Smolin (see Drees 1990, chapter 2; Drees 1993; see also Isham 1993). Let me briefly characterize these three approaches.

For Stephen Hawking (and similarly more recently Julian Barbour), reality deep down might be timeless. Time is just a parameter which may have a finite past, but nothing extraordinary is there to be said about '$t = 0$'. For a particular choice of parameters it is a boundary, but other states of affairs would be the boundary if the parameter were defined differently.

For Alexander Vilenkin and Roger Penrose, there is something remarkable about the initial state, in need of an adequate description in the physics. They had come up with particular proposals on the Weyl curvature (Penrose) and on the boundary conditions for the wave function of the Universe (Vilenkin) which seek to articulate the remarkable properties of the initial state as a consequence of a fundamental rule specifying the structure of reality. Andrei Linde, and differently Lee Smolin, give time an even more prominent role, by arguing that the specific features of the early universe are not a consequence of a fundamental rule, but rather of a preceding process; the initial conditions of the observable universe are product of history, not of law. Smolin has suggested a cosmic 'darwinism' in which universes may have daughters (and granddaughters, and so on), with the universe that has the best conditions for generating such daughters becoming the most frequent member of the whole set of universes, and thus the most likely one to be found as our universe. Others put more stress on the chaotic character of the underlying plurality of 'universes', with a role for dissipative processes to wipe out specifics of its initial state.

This extremely brief survey indicates what Jeremy Butterfield and Christopher Isham (2001, 38) wrote about theory construction in quantum gravity.

> In this predicament, theory-construction inevitably becomes much more strongly influenced by broad theoretical considerations, than in mainstream areas of physics. More precisely, it tends to be based on various *prima facie* views about what the theory *should* look like – these being grounded partly on the philosophical prejudices of the researcher concerned, and partly on the existence of mathematical techniques that have been successful in what are deemed (perhaps erroneously) to be closely related areas of theoretical physics, such as non-abelian gauge theories. In such circumstances, the goal of a research programme tends towards the construction of abstract theoretical schemes that are compatible with some preconceived conceptual framework, and are internally consistent in a mathematical sense. The situation … tends to produce schemes based on a wide range of philosophical motivations, which (since they are rarely articulated) might be presumed to be unconscious projections of the chtonic psyche of the individual researcher – and might be dismissed as such! Indeed, practitioners of a given research programme frequently have difficulty in understanding, or ascribing validity to, what members of a rival programme are trying to do. This is one reason why it is important to uncover as many as possible of the assumptions that lie behind each approach: one person's 'deep' problem may seem irrelevant to another, simply because the starting positions are so different.

Changes in conceptuality have been typical of fundamental transitions in physics, such as those from classical physics to quantum physics and from Newtonian conceptions of space and time to those of the special and general theories of relativity. With the development of new scientific theories, our knowledge of the world was not merely enlarged to include the very small, large or fast. In many ways, our knowledcge was restructured, with entities and structures postulated that had not been envisaged before. New theories led to a reinterpretation of the world. With respect to practical, observable and instrumental aspects the old theory is a continuous limiting case of the new one, but conceptually or ontologically it is radically different. Newton's law of gravity can still be used for almost all practical purposes, even though the conceptuality of the better theory, General Relativity, is different. Empirical or observational consequences of previous theories, as far as corroborated by experiments, must be reproduced by a new theory, even if the new theory is cast in radically different conceptions. Such a transition is at stake with respect to quantum cosmology as well: it leads to a reinterpretation of our concepts regarding the world, and especially the concept of 'time'. And such a change is not restricted to cosmology, as the theory at stake is quantum gravity which would be a more fundamental replacement of Newtonian and Einsteinian views of space and time. If such a radical change in our ontological conceptuality is possible, due to the need to replace General Relativity by a quantum theory of spacetime, all aspects of the universe including its temporal nature, is open for reinterpretation, and not merely the absoluteness of the apparent beginning.

Beyond a First Cause?

When we realize such limitations of the Big Bang theory, the question may be not what might be 'beyond the Big Bang?' but rather what will come when we consider future theories. It may have more popular appeal to ask what would be 'beyond the Big Bang'; in the context of the cosmological argument, the answer would be expected to be the Creator, lighting the fuse as a great engineer. It probably was this popular appeal that made a publisher decide that the title of a book of mine was to be *Beyond the Big Bang* (Drees 1990), even though I had offered the manuscript with the title *Beyond the Big Bang Theory*, the point being that future theories might be conceptually quite different. One option might be that time is just a parameter. In this case, the special characteristics of 'the Big Bang' would be merely a matter of description, just as the 'North Pole' is on many maps. Another option, much discussed as if it were an alternative to an understanding of the Big Bang as the moment of creation, is the idea that our observable universe is just one domain within a much larger reality. Such ideas are called 'multiple universes', or a 'multi-verse'.

If one thinks of other domains aside of our observable universe, the Big Bang would be more like the beginning of one individual life than of reality as such. In such conceptual schemes our observable universe would be a domain or epoch within a larger framework. (It is hard to avoid spatial or temporal notions, but 'domain' could also be in some other conceptualization. Perhaps we should envisage multiple 'temporalities' in parallel. Some colleagues in 'religion and science' dismiss speculative theories about 'multiple universes', but I am not that skeptical.

One reason to take such an expansion of the idea of reality serious is history. Once the geocentric universe was replaced by a heliocentric one, it was not that big a step to think of stars as other 'suns', each with their own set of planets. And much later a discussion arose whether certain nebulous spots where like clouds within our Galaxy, or whether they were 'island universes' by themselves. It turned out to be the case that some indeed were – but by then the grand name 'island universes' was replaced by galaxies. So too, I think, when one speaks of 'multiple universes' or a 'multi-verse'. The concept of a 'universe' does not easily allow for a plural – but one can easily imagine multiple domains (including our observable environment) in a larger framework that would then be the universe, larger than thought of before.

There might be theoretical reasons to assume that there are multiple domains. If a theory consistent with current observations would predict the formation of other domains alongside ours, it would be *ad hoc* to exclude those domains. Before other planetary systems were observed, we had good reason to assume that there would be planetary systems with some other stars, as any reasonable explanation of the formation of planetary systems made it something that could happen more than once. Claiming uniqueness because we don't observe other domains, may well be premature. Corroboration by observations is not to be excluded, as future observations within our domain may show traces of earlier domains or 'interaction effects' with neighboring domains allowed by the theory.

A theory that predicts 'many universes' (better 'many domains'), might resolve some scientific questions about the specific properties of our cosmic environment. However, the fundamental questions have not disappeared. The larger system too happens to exist – and is thus open to the question why it exists. The mystery of existence has shifted to a more encompassing system, but hasn't disappeared. The larger question remains – again and again at the boundary of science.

The Persistence of the Mystery of Existence

If a quantum cosmology programme would be successful in embedding the Big Bang cosmology in a larger framework that integrates all known physics in a coherent and consistent whole, would this be the end of the cosmological argument? Or might it be the end of a particular form of the cosmological argument, namely one that depends on a beginning of time – but not of other forms, that reflect upon the contingency of existence, expressed in the question 'Why is there something rather than nothing?'

Science is successful; much is explained or might be explained with the progress of the various disciplines. Will understanding ever be exhaustive, or is there a remnant of unanswered questions – perhaps not a question about a 'first cause', but of a similar kind? I am convinced that there always are such questions, though not always the same as knowledge advances. Basically, scientists find traces and clues – and seek to understand the past or the inner workings of organisms or galaxies. In that process, we answer questions and pass on other questions. There is a huge division of labour.

An architect who designs a building using concrete. He may have knowledge of the forces that this concrete will be able to withstand. If asked why the strength is as it is, the architect might refer us to an engineer who studies material sciences. This engineer should be able to inform us about experiments and the relevant theory, about the wear and tear of the materials concerned, and their relations to chemical bonds between the various materials. Perhaps the engineer even knows from which geological deposit the sand and cement have been taken. However, if you go on asking how those layers came to be there, the engineer will refer to a geologist. The geologist can tell a story about the erosion of mountains and sedimentation of sand and stones by rivers. Perhaps the geologist can discover that the sand used was part of a particular mountain range, and perhaps even that the same material was already deposited on a sea floor before. However, if one continues by asking where the silicon and oxygen come from, the chemical elements making up sand, the geologist will have to say that these were there when the Earth formed. For further questions, he will refer to the astrophysicist. And the astrophysicist speak about the formation of elements out of hydrogen in the interiors of stars and during supernova explosions, and the way these elements are distributed in the universe and may get included when a solar system forms. However, this explanation assumes that there is already hydrogen as the material out of which stars are formed. When we go on with 'historical' questions we come to theories about the earliest stages of the universe, to the turf of the cosmologist.

This, in a nutshell, is typical of science. Scientists answer questions belonging to their domain of expertise, while passing on other questions, about the things they take for granted in their own work. (The image of passing on questions was developed in 1868 by Thomas Huxley in a lecture 'On a piece of chalk', arguing for the coherence of various explanatory angles, and re-used by Steven Weinberg in his *Dreams of a Final Theory*. Even if our current questions are answered by future theories, new questions emerge in the context of those future theories. For instance, the inflationary model in cosmology solves many questions regarding the universe as observed, but does not explain why the universe is such that inflation happens; some assumptions are always made. The reach of explanation is impressive, but explanatory successes do not exclude further questions. Again and again, questions emerge at the limits of scientific understanding. They may be resolved, but the deeper understanding will have other elements that are assumed rather than explained.

Questions remain even if physics and cosmology agree one day on a theory explaining all known phenomena in a unified, coherent way. Imagine, a single article, a single formula answering all our questions. But the article is on a piece of paper; the formula consists of symbols. Thus, there is no answer to the question: Why does reality behave as described here? It is as with a drawing of the Belgian artist René Magritte. It is a careful drawing of a pipe, a pipe used for smoking tobacco. Underneath it, he had written 'Ceçi n'est pas une pipe' – 'This is not a pipe'. And he was right. It is an image of a pipe. One cannot fill the image with tobacco and if one would attempt to light the image, something else happens than when one lights a pipe. There is a difference between an image, how accurate it may be, and reality. This is also the case for a good scientific theory. However accurate the theory, the question remains why reality behaves as it does, and as described in the theory.

There is a traditional philosophical question: Why is there something rather than nothing? And there are similar philosophical questions that arise due to science, but are not answered by science. Why is mathematics so effective in describing reality? Why is reality such that we can work well with wrong, or at least incomplete theories? For this is our predicament. It is a mistake to inflate problems and puzzles to mysteries, which would perhaps only be open to a religious answer. Such an approach would be forced into further retreats again and again. However, the success of science in solving puzzles and problems can itself evoke questions. How can human science be so successful? What does that say about humans and about reality?

There are various ways of dealing with such persistent questions. It is said that the American president Truman had a sign on his desk saying 'The buck stops here'. In a company or administration one can pass on hard decisions to persons higher up, but the president cannot avoid responsibility. Scientists, however, do not have to make a choice. They may have to live with the insecurity of unanswered questions. A political decision or dogmatic answer is neither necessary nor adequate. Religious people do not have to cut this Gordian knot either. They ought to be willing to recognize that our explanatory quest is open ended. The physicist Charles Misner (1977, 97) expressed this well:

> Saying that God created the universe does not explain either God or the Universe, but it keeps our consciousness alive to mysteries of awesome majesty that we might otherwise ignore, and that deserve our respect.

In my opinion, this dynamic of questions and answers be understood as a natural variety of transcendence. This would be an epistemic type of transcendence; the progress of theories does not bring one to an ontologically different level. May be some of the philosophical questions that seem to be independent of theoretical progress are different, as they do not so much address a horizon that shifts but rather a more generic question – but then one gets into the philosophical complexities of the philosophical 'cosmological argument'.

The persistence of questions, may lead some to a sense of gratitude and wonder about the existence of our world. This wonder or puzzlement about the contingency of existence, and its order and intelligibility, is something that receives an answer of some sort from faith in a transcendent God who endows the world with existence and order. However, this move would not be from science to faith, as if the postulate that such a God exists were the conclusion of an inference to the best explanation of the natural world.

And perhaps, the transcendent – if such be – is to be thought of differently, not as the next answer in a sequence of answers and questions – as one may always ask how God came into existence. Given the role of mathematics, and its independence from the physical dimensions of time and space, one might try to draw on mathematics and logic to imagine 'transcendence'. Axioms are not so much the cause of the theorems, but rather the formal ground of all subsequent theorems.

4. God as Ground?

'Transcendence' most often intends to refer to God, a divine being 'up there', distinct from the natural and human world. Given that the spatial metaphor implicit in the common sense meaning of 'transcendence' seems hard to hold onto in our understanding of the cosmos, how else could the term be understood?

Mathematics

Mathematics is odd if one comes at it from an empiricist mind set. Pure circles, triangles, cubes and the like do not exist, nor do imaginary numbers, Lie groups or Bessel functions. Nonetheless, we can make well-defined claims about their properties, and argue about the truth or falsity of various mathematical claims. We can even make mathematical existence claims such as that there is (or that there is not) an even number that is not the sum of two primes (Goldbach's conjecture). The fact that we currently don't know which option is right, doesn't undermine the conviction that either there is such a number or there isn't one – and that the truth is not dependent upon human preferences.

One interpretation of this feature of mathematics has been 'Platonism' (used here without any regard for historical accuracy), the view that mathematical realities exist 'out there' in an objective but immaterial world. Thus, mathematical truth can be understood as a form of correspondence between our propositions and mathematical reality. Mathematicians explore a pre-existing world, and make discoveries. Roger Penrose (1989) seems a contemporary advocate of such a view.

As an ontology this 'Platonic' reality is so distinct from material reality that it is hard to envisage 'where' it might be. And if one dismisses this as a non-problem, given the categorical difference between material reality and this Platonic reality, a second problem arises: how do we, material beings, have access to those non-material lands? 'Mathematical intuition', the possibility to make 'observations' in this Platonic realm, would be a remarkable addition to the experiential, causally mediated repertoire we are supposed to have. Such an ontology of mathematics seems too remote to fit the epistemic challenge how mathematical knowledge is acquired and developed (e.g., Kitcher 1984, 102). A different but somewhat related problem is how it might be possible that mathematics is useful for the physical world, if it is categorically distinct, dealing with abstract entities rather than material objects and natural processes.

A quite different view of the nature of mathematics is constructivist in kind. Leopold Kronecker is supposed to have said that God made the natural numbers (1, 2, 3 ...); the rest is the work of humans – 'Die ganzen Zahlen hat der liebe Gott gemacht, alles andere ist Menschenwerk'. Mathematical objects are human creations, a conceptual world that is 'up to us'. But then, if it is just our construction, why don't we see much more variation? Why do mathematicians agree on mathematical insights, across cultural, linguistic and ideological borders? If we would ever encounter extraterrestrial mathematicians, we should expect them to have a different notation, but fundamentally the same mathematics. Can constructivism do justice to the universality of mathematical insights?

Platonic and constructivist approaches offer ways of understanding transcendence. The Platonic view has a transcendent reality of abstract entities, ready to be explored. A constructivist lacks such a transcendent ontology, but our experience with mathematics as universal might be interpreted as suggesting that there is some a form of transcendence in the process of mathematical abstraction.

Let us go back to the beginnings of human mathematics. The basics of counting and measuring are clearly to be found in the practical life of early hominids, as counting is useful when sharing food and keeping track of enemies, and in trade and agriculture. One of the simplest instances of Pythagoras' formula, namely $3^2 + 4^2 = 5^2$ has been discovered early in human history as a way to create effectively straight corners. The general form came later, through abstraction and reflection. It would be this process of abstraction and reflection that goes well beyond the concrete natural situation in which it has its beginnings, that generates mathematical insights. A remarkable feature of our reality is that those abstract insights turn out to be applicable to many other situations as well. Apparently, natural processes have many features that are structurally the same.

Thus, though not endowed with a distinct type of perception or intuition (as needed upon a Platonic view), humans can come up with genuine mathematical knowledge developed by abstraction, generalization and idealization from the natural. Transcendence would not be about a ontologically distinct realm, but rather reflect the possibility of abstraction from concrete processes to structures that in their abstraction are 'universal' in kind, and the related play of imagination that allows for mathematical constructs that have no straightforward real world applicability.

The philosophy of mathematics is a scholarly profession in itself, which I will not delve into here, but this analogy serves as an indication that one need not abandon a naturalistic understanding of human behavior in practice (counting, measuring) to consider the possibility of universal concepts that go beyond specific situations, and in that sense are not factual but ideal or transcendental. Perhaps something similar might be argued for morality – arising out of human practices serving self-interest through cooperation, but in the process of reflection and justification reaching for more abstract and apparently universal values.

The View from Eternity

The religious vocabulary associated with such transcendence is considered in the work of Stewart Sutherland, *God, Jesus and Belief: The Legacy of Theism* (1984) as the view *sub specie aeternitatis*, that is, the view from the perspective that is not a particular perspective, and thus not serving a particular self-interest. (And any self-interest is, by definition, particular). Thus, in a novel someone contemplating a modest job as a school teacher, is told that if he does it well, 'You will know it, your pupils will know it; and God will know it' (Sutherland 1984, 87). It is the 'God will know it', that lifts the considered course of action to a higher plain. Sutherland argues that it is not accidental that the language of theism is used. 'The language of theism embodies, offers and protects the possibility of a view of human affairs *sub specie aeternitatis*' (Sutherland 1984, 88). He points to two beliefs or hopes involved, namely that one may transcend the particulars of an individual, community or age, and even that the ultimate context in which our behavior is to be judged is against values that are beyond 'the outlook of mankind' (88) and particulars of the species.

Such elements are somewhat reminiscent of the universals of mathematics, though there the ability to build consensus among those with expertise is far greater than in the moral domain, where cultural and individual differences are more common and persistent. (By the way, the analogy with math is also considered by Sutherland (1984, 91)).

Anselm's Ontological Argument and Findlay's Objection

Early in this contribution, we considered 'natural tyheology', with the argument from design and the cosmological argument. There is also another type of theistic argument, that is fairly similar to arguments in mathematics, namely the so called 'ontological argument'. The classic example is the argument formulated in the 11th century by Anselm of Canterburry in his *Proslogion*. God is to be thought of as the being greater (more perfect) than which nothing can be thought to exist. If this God did not exist, we could think of something greater – namely this God with real existence. Hence, God must exist. Another version makes the kind of existence special: whereas all created entities have contingent existence, God has necessary existence – it is unimaginable that God would come into existence or go out of existence.

In making the case on the basis of a definition and logical reasoning, this argument has some similarity with mathematics. Is mathematics promising as a model for a theistic understanding of transcendence as pointing to God's existence as real and different from the world of creatures? It may seem attractive that the notion of transcendence can be understood in terms of mathematics, but there is a downside: the argument might be understood more appropriately as an argument for the non-existence of God. In an article on arguments for the existence of God, with special consideration of Anselm's ontological argument, J.N. Findlay (1955) has pointed out that precisely because of the nature of the argument, God is not placed with (contingent, empirically real) entities that have genuine existence but rather with mathematical entities and other conceptual truths, that have no claim to existence. As Findlay (1955, 54) wrote:

> It was indeed an ill day for Anselm when he hit upon his famous proof. For on that day he not only laid bare something that is of the essence of an adequate religious object, but also something that entails its necessary non-existence.

This regards mathematical truth. There is no mathematical triangle in physical reality, though there are plenty of objects that approximate a mathematical triangle. But even if there is no triangle, it is true that all triangles have three angles (and that in an Euclidean space these add up to a 180 degrees, et cetera). Precisely because one can suspend the question whether triangles exist, one may have mathematical conclusions of universal validity. Findlay's challenge also regards the moral perspective advocated by Sutherland – precisely because viewing our choices *sub specie aeternitatis* is a perspective that could be said not to exist or be available, might it function as a major regulative notion.

Thus, abstraction is a mixed blessing in the present context. On the one hand, the move to transcendence in mathematical terms, rather than in temporal ones, frees one from the challenges that arise as the concept of time changes or even disappears from the fundamental theories. On the other hand, the kind of existence becomes more abstract, far removed from existence in the sense of our experiences with objects we perceive.

God as Ground?

Perhaps we should take even more license from our concepts of time, space, and cause, and from God as an entity that exists, separately from the existence of empirical reality. Speaking of God as Ground of Being softens somewhat the dualistic scheme of God and creation, but does not fundamentally undermine it. A major figure in the articulation of such a theological position has been Paul Tillich (Wildman 2006). This view has come to be formulated often in panentheistic terms: understanding the world to be in God, even though God surpasses the world (Clayton and Peacocke 2004). I found a most inspiring poetic expression of this view among aphorisms in *The Aristos* of the novelist John Fowles (1980, 27):

> The white paper that contains a drawing; the space that contains a building; the silence that contains a sonata; the passage of time that prevents a sensation or object continuing forever; all these are 'God'.

5. The nature of theology: not just natural theology

We started this essay with natural theology, and its quest to build arguments for the existence of God on our knowledge of nature. However, not all theology is natural theology. We will consider here two other approaches in theology.

Schleiermacher: Religion is Different from Cosmology

As a quite different voice let us consider Friedrich Schleiermacher, a German theologian. In the second of his speeches on religion 'to its cultured despisers', from 1799, he recognized that metaphysics, morality, and religion all deal with 'the universe and the relationship of humanity to it. This similarity has long since been a basis of manifold aberrations; metaphysics and morals have therefore invaded religion on many occasions, and much that belongs to religion has concealed itself in metaphysics or morals under an unseemly form' (Schleiermacher ([1799] 1996, 19)). He acknowledges that all three have the same subject matter, namely reality (the universe) and the relationship of humanity to it. However, Schleiermacher (1996, 20) is very critical of carrying over notions from one side to the other.

> You take the idea of the good and carry it into metaphysics as the natural law of an unlimited and plenteous being, and you take the idea of a primal being from metaphysics and carry it into morality so that this great work should not remain anonymous, but so that the picture of the lawgiver might be engraved at the front of so splendid

> a code. But mix and stir as you will, these never go together; you play an empty game with materials that are not suited to each other. You always retain only metaphysics and morals. This mixture of opinions about the highest being or the world and of precepts for a human life (or even for two) you call religion! (...) But how then do you come to regard a mere compilation, an anthology for beginners, as an integral work, as an individual with its own origin and power?

Thus, integration is premature. For Schleiermacher (22f), the resolution has been for religion to take leave from any engagement with metaphysics and morals.

> In order to take possession of its own domain, religion renounces herewith all claims to whatever belongs to those others and gives back everything that has been forced upon it. It does not wish to determine and explain the universe according to its nature as does metaphysics; it does not desire to continue the universe's development and perfect it by the power of freedom and the divine free choice of a human being as does morals. Religion's essence is neither thinking nor acting, but intuition and feeling. It wishes to intuit the universe, wishes devoutly to overhear the universe's own manifestations and actions, longs to be grasped and filled by the universe's immediate influences in childlike passivity. Thus, religion is opposed to these two in everything that makes up its essence and in everything that characterizes its effects.

In the tradition of natural theology there has always been a strong interest in cosmological and other scientific knowledge as evidence of the well-designed character of our world. However, Schleiermacher's faith is more existential in orientation: it deals not with facts or theories, but with the way we relate to reality. For the Christian, this is our sense of creatureliness, of 'absolute dependence' as Schleiermacher calls it.

Another Christian Tradition

Let me point out a third position, different from natural theology but not as indifferent to knowledge of the world as Schleiermacher. This strand is represented by Augustine, theologian of the early church; Thomas Aquinas, a major thinker of the European Middle Ages; and many others. These thinkers combined a strong theological agenda with appreciation of the natural philosophy of their day. Two features seem to distinguish philosohical-theological reflections in this tradition.

One is respect for the multifacetted character of the Biblical texts, and thus awareness of the need for interpretation. If Scripture and science conflict, it may be better to say that *our interpretation* of Scripture conflicts with science – and thus, there might be need for a different understanding of Scripture (e.g., McMullin 2011).

The other is the reflective engagement with the best available natural philosophy (anachronistically one might say 'science'). A recent representative of this style of thinking has been the historian and philosopher and historian of science Ernan McMullin, who wrote the following at the end of an exemplary article on the question 'How Should Cosmology Relate to Theology?':

> The Christian cannot separate his science from his theology as though they were in principle incapable of interrelation. On the other hand, he has learned to distrust the simpler pathways from one to the other. He has to aim at some sort of coherence of world-view, a coherence to which science and theology, and indeed many other sorts of human construction like history, politics, and literature, must contribute. He may, indeed, *must* strive to make his theology and his cosmology consonant in the contributions they make to this world-view. But this consonance (as history shows) is a tentative relation, constantly under scrutiny, in constant slight shift. (McMullin 1981, 52)

Among these philosophical-theological thinkers, quite a few have stressed the categorical difference between God the Creator and all creatures, a difference that is articulated by understanding God as timeless (that is, as someone to whom temporal distinctions do not apply). With such a view of God, many issues regarding the beginning of the world, its continuity (sustenance), and its dynamics acquire a particular shape. The result is quite different from the design arguments in the tradition of natural theology.

So far, we have considered modern cosmology and three different types of response that have emerged in Western Christianity – a religious appropriation in the tradition of natural theology, the friendly distinctiveness of Schleiermacher, and a mutual engagement without dependence. It should be clear by these three examples that there is diversity within a single tradition, and that this diversity also reflects different ideas of what religion and theology should be.

6. Acknowledgements

In this chapter, I have liberally re-used elements and passages from other writings such as Drees 1990, 1993, 1996, 2002, 2010, 2012.

References

Augustine Aurelius [400], *The Confessions of Saint Augustine*, translated, with an introduction and notes, by J.K. Ryan, Doubleday, New York 1960.

Brooke J.H., Cantor G., *Reconstructing Nature: The Engagement of Science and Religion*, T&T Clark, Edinburgh 1998.

Butterfield J., Isham Ch.J., *Spacetime and the philosophical challenge of quantum gravity*, [in:] eds. C. Callender, N. Huggett, *Physics Meets Philosophy at the Planck Scale: Contemporary Theories in Quantum Gravity*, Cambridge University Press, Cambridge 2001, pp. 33–89.

Clayton Ph., Peacocke A., eds., *In Whom We Live and Move and Have Our Being: Panentheistic Reflections on God's Presence in a Scientific World*, Eerdmans, Grand Rapids 2004.

Drees W.B., *Beyond the Big Bang: Quantum Cosmologies and God*, Open Court, La Salle 1990.

Drees W.B., *A case against temporal critical realism consequences of quantum cosmology for theology*, [in:] eds. R.J. Russell, N. Murphy, C.J. Isham, *Quantum Cosmology and the Laws of Nature: Divine Action in Scientific Perspective*, Vatican Observatory Press, Center for Theology and the Natural Sciences, Vatican City State and Berkeley, CA 1993.

Drees W.B., *Religion, Science, and Naturalism*, Cambridge University Press, Cambridge 1996.

Drees W.B., *Creation: From Nothing until Now*, Routledge, London 2002.

Drees W.B., *Religion and Science in Context: A Guide to the Debates*, Routledge, London 2010.

Drees W.B., *Human insignificance? Cosmology and creation stories*, [in:] *Science and the World's Religions. Volume 1: Origins and Destinies*, eds. P. McNamara, W.J. Wildman, Praeger, Santa Barbara, CA 2012.

Findlay J.N., *Can God's existence be disproved?* [in:] *New Essays in Philosophical Theology*, eds. A. Flew, A. MacIntyre, SCM, London 1955, pp. 47–56.

Fowles J., *The Aristos*, Triad/Granada, Rev. ed. Falmouth 1980.

Harrison P., *Religion, the royal society, and the rise of science*, "Theology and Science" 2008, no. 6 (3), pp. 255–271.

Huxley Th.H., *On a piece of chalk*, "Macmillan's Magazine" 1868. Reprinted in T.H. Huxley, *Collected Works, Volume VIII. Discourses: Biological & Geological*, Macmillan, London 1894, pp. 1–36.

Isham Ch.J., *Quantum theories of the creation of the universe*, [in:] *Quantum Cosmologies and the Laws of Nature: Scientific Perspectives on Divine Action*, eds. R.J. Russell, N. Murphy, C.J. Isham, Vatican Observatory Publications & Berkeley: Center for Theology and the Natural Sciences, Vatican City State 1993, pp. 49–89.

Kitcher Ph., *The Nature of Mathematical Knowledge*, Oxford University Press, Oxford 1984.

McMullin E., *How should cosmology relate to theology?* [in:] *The Sciences and Theology in the Twentieth Century*, ed. A.R. Peacocke, Oriel Press & Notre Dame: University of Notre Dame Press, Stocksfield 1981.

McMullin E., *Darwin and the other christian tradition*, "Zygon: Journal of Religion and Science" 2011, no. 46, pp. 291–316.

Manson N.A., ed. *God and Design: The Teleological Argument and Modern Science*, Routledge, London 2003.

Misner Ch.W., *Cosmology and theology*, [in:] *Cosmology, History, and Theology*, eds. W. Yourgrau, A.D. Breck, Plenum Press, New York 1977.

Numbers R.L., *The Creationists: The Evolution of Scientific Creationism,* University of Chicago Press, Chicago.

Penrose R., *The Emperor's New Mind: Concerning Computers, Minds, and the Laws of Physics*, Oxford University Press, Oxford 1989.

Schleiermacher F., *On Religion: Speeches to its Cultured Despisers*, transl. R. Crouter, Cambridge University Press, Cambridge [1799] 1996. Translation of *Über Religion: Reden an die Gebildeten unter ihren Verächtern*.

Sutherland S.R., *God, Jesus & Belief: The Legacy of Theism*, Basil Blackwell, Oxford 1984.

Weinberg S., *The First Three Minutes*, Basic Books, New York 1977.

Weinberg S., *Dreams of a Final Theory*, Pantheon Books, New York 1992.

Wildman W.J., *Ground-of-being theologies*, [in:] *The Oxford Handbook of Religion and Science*, eds. Ph. Clayton, Z. Simpson, Oxford University Press, Oxford 2006, pp. 612–632.

Index of Names